G.A.I.N.S.

Games, Activities, Ideas
for Number Sense

Dr. Carl H. Selzer

Didax
Educational Resources

Order number 2-149
ISBN 1-58324-172-8

A B C D E F 05 04 03

Didax
Educational Resources
395 Main Street
Rowley, MA 01969
www.didaxinc.com

Contents

Understanding Operations

Large Numbers

Games

Answers

Introduction

Number sense is an ability that is difficult to define. The presence or lack of number sense in students is usually very obvious. My definition of number sense includes:

1. understanding the meaning of numbers and operations with numbers

2. being able to determine how certain numbers are alike and how they are different

3. being able to round off numbers

4. being able to estimate the results of mathematical operations

5. being able to determine whether an answer is reasonable or unreasonable

6. having a sense of a number's value in relation to other numbers

7. understanding the compatibility of given number sets

8. being able to understand conventional algorithms and to invent others

9. being able to apply basic mathematical principles

10. being able to see patterns and relationships between numbers

When students possess number sense, they are able to recognize patterns and relationships between numbers. They can take numbers apart and put them back together in various configurations and comprehend positionality of number sets. Number sense leads to a greater ability to solve more complicated problems. Students with number sense can solve many mathematical problems mentally. Finally, students who possess number sense find math more satisfying.

Over the past ten years of teaching, I have become convinced that Dr. Constance Kamii is correct in stating that the teaching of traditional algorithms is harmful to a child's ability to think mathematically. When children do "mental math" using number sense concepts, they are required to *think* the problem through, rather than simply following a set of rules. All the activities in this book are designed to be done without a calculator. We want children to be able to feel comfortable with the symbolic representations of numbers themselves. Having this comfort level makes it much easier to apply mathematics to solving real-life problems.

Mathematics is the science of patterns and relationships. Every scientist engages in four basic activities:

1. They conduct experiments, investigations, and explorations;

2. They observe their experiments;

3. They draw conclusions and form hypotheses;

4. They verify that what they think is happening is indeed true.

This to me is a lesson plan for my classroom. I want as much as possible to have these four components present in any lesson. Discovery is an important motivator. Let your students discover things for themselves; don't be too quick to give answers.

G.A.I.N.S.: Games, Activities, Ideas for Number Sense, includes the following basic ideas and activities to promote number sense:

- Finding Patterns in Number Sets
- Place Value
- Estimating
- Reasoning and Proof
- Understanding Operations
- Large Numbers
- Games

The reproducible exercises in this book can be used in any sequence, but in general, the exercises within each section build on the previous exercises in that section. The instructions to the student are as brief as possible as we want them to concentrate on discovering why numbers interact to produce specific results. There are ample opportunities for students to write about their observations and hypotheses. You may want to develop some additional questions and writing activities as you become comfortable with the exercises.

Finding Patterns in Number Sets
• • • • • • • • • • • • • • • •

See pages 2–13

These exercises are designed to help children recognize, analyze, extend, and describe patterns represented with symbols. You may want to encourage children to analyze the structure of simple patterns so they can generalize to the more complex patterns included in these exercises. Writing about the patterns is at least as important as solving them; there are a number of ways to analyze the structure of each pattern. You can have the children share their discoveries with classmates.

Here is an example (**write the following on the board**):

1, 4, 9, 16, 25, 36, ___, ___, ___

Fill in the three blanks with the correct answers: 49, 64, 81.

Here are two appropriate analyses of the pattern you can explain:

Student 1: This is a list of squares of whole numbers from 1 to 9 in order.

Student 2: $a + 3$, $a + 5$, $a + 7$, $a + 9$, etc. Always goes up by two more.

In general, children should be able to move from considering individual mathematical objects (in this case symbols) to thinking about classes of objects.

Place Value
• • • • • • • • • • • •

(See pages 14–17)

Understanding place value is more than knowing that single-digit number increment to double-digit numbers, to triple-digit numbers, etc. The exercises are designed to help children understand the multiplicative nature of our number system.

Estimating
· · · · · · · · · ·

(See pages 18–30)

Explain to the students that sometimes when we estimate we round off numbers in a way that is not consistent with a specific rule, but seems to be the most logical way.

In the example below (**write the following example on the board**):

$$1,359 \div 130$$

we might round to the nearest 100 as 1,400. One could also say:

$$1,359 \text{ rounds to } 1,300 \div 130 = 10,$$

which is just as reasonable and may be just as easy to do. Actually,

$$1,359 \div 130 = 10.45.$$

Another example is (**write the following example on the board**):

$$138 \div 6 = \frac{120}{6} + \frac{18}{6}$$
$$= 20 + 3$$
$$= 23$$
$$249 \div 7 = \frac{210}{7} + \frac{1}{7}$$
$$= 30 + \frac{35}{7} + \frac{4}{7}$$
$$= 35 + \frac{4}{7}$$

It is important to think carefully about how we are estimating. Since we want to calculate mentally, we want the calculations to be made as easily as possible! For this reason students need to recognize compatible numbers and be able to deal effectively with them.

Explain to the class that compatible numbers are numbers that "go together nicely." An example would be multiples of 10, such as: 10, 100, 1,000, and 10,000. One could expand the idea to multiples of 5, such as: 5, 10, 15, 20, and 25.

Other examples of compatible numbers might include:

$$3 + 7, 25 + 15, \text{ and } 150 + 250.$$

These are numbers that can help us to estimate calculations more easily.

For example (**write the following example on the board**):

$$307 - 205$$

could be calculated mentally by changing it to:

$$305 - 205 = 100$$

and, since $307 - 205$ would be 2 more, the correct answer is 102. We could also say:

$$307 - 200 = 107$$

but 200 is 5 less than 205, so we need to subtract 5, or

$$107 - 5 = 102.$$

Experiencing G.A.I.N.S.

Another example (**write the following example on the board**):

568 − 159

could be changed to:

469 − 159 = 310

469 − 1 − 159 = (469 − 159) − 1 = 309.

A third example (**write the following example on the board**):

564 − 119 = ?

An easier calculation would be:

564 − 120

but the difference between these two numbers would be 1 less than 564 − 119. Therefore, we need to add 1 back. So:

564 − 120 = 444

Therefore:

564 − 119 = 444 + 1 = 445

A fourth example (**write the following example on the board**):

805 − 426 = ?

An easier calculation would be:

800 − 426 = 374

But:

805 − 426

is 5 more, so we need to add

374 + 5 = 379.

These four examples are meant to illustrate possible approaches to solving these problems mentally.

Students may come up with several different ways of solving the problems in the following exercises. It is more important that they demonstrate reasonable estimating skills rather than finding the exact correct answer.

Reasoning and Proof
• • • • • • • • • • • • • • • • • • • •

(See pages 32–51)

Mathematical reasoning is developed when children are encouraged to express their own ideas for examination. There are a number of ways to arrive at the same result and the exercises in this section are designed to reinforce this. "**Now That Makes Sense!**" should generate some lively discussions as children justify their thinking about the magnitude of their answers. Encourage children to expand on their reasons.

It is key for children to have an idea about what constitutes a convincing argument. For example (**write this on the board**):

A person could skateboard 300 miles in a day.

Describe the following thought process:

"There are 24 hours in a day, but a person could only be on a skateboard for about 8 hours at the most. A car can travel at about 50 miles per hour on a highway, but skateboards can only go on sidewalks. So maybe a skateboard can travel at 5 miles per hour. So 5 x 8 is only 40, so 300 miles is *not* a reasonable answer."

Understanding Operations
• •

(See pages 52–64)

Many students will be comfortable with using the basic operations, but combining these operations to solve problems requires an understanding of properties of operations, such as the distributive property.

The distributive property can help to solve multiplication problems and to improve computational fluency by breaking unfamiliar numbers into "compatible" numbers. For example **(write this on the board)**:

$9 \times 17 = ?$

This can be made easier by distributing in two ways:

$9 (10 + 7) = 90 + 63 = 153$, or

$9 (8 + 9) = 72 + 81 = 153$

Addition problems can be solved using the associative property. That is:

$(1 + 2) + 3$

is the same as:

$1 + (2 + 3)$.

Point out that only the times tables to ten were used to solve this problem.

Large Numbers

(page 66–77)

To compute fluently, children need to be able to use operations to mentally solve problems with large numbers. These exercises use squared numbers to demonstrate the useful patterns that can be applied to solving multiplication problems with large numbers.

NCTM STANDARDS

	Finding Patterns Pages 2–13	Place Value Pages 14–17	Estimating Pages 18–30	Reasoning and Proof Pages 32–51	Understanding Operations Pages 52–64	Large Numbers Pages 66–77	Games Pages 81–92
Number and Operations							
Place Value		X		X			X
Equivalent Representations			X	X			
Fractions	X		X	X	X		X
Factors	X		X	X	X	X	X
Multiplication and Division	X		X	X	X	X	X
Relationships between Operations					X		
Properties of Operations					X	X	X
Fluency with Operations	X	X	X	X	X	X	X
Using Mental Math	X	X	X	X	X	X	
Estimation			X	X	X	X	X
Selecting Appropriate Methods			X	X	X		X
Algebra							
Recognize and Extend Patterns	X			X	X	X	
Associativity and Distributivity	X	X	X	X	X	X	X
Using Equations					X		
Analyzing Changes in Variables	X	X	X	X	X	X	X
Data Analysis and Probability							
Collecting Data							X
Predicting Outcomes			X	X			X
Likely or Unlikely				X			X
Problem Solving	X		X	X	X	X	X
Reasoning and Proof				X	X		X
Communication	X		X	X	X	X	X

FISHBURGER PALACE

LESSONS

Find the Pattern I

Look at the sets of numbers. Find the pattern for each set and fill in the blanks. What is the pattern? Write it on the line provided.

Example:

1. 2, 4, 6, 8, 10, 12, 14, _16_, _18_, _20_, _22_, . . .

Even numbers, starting with 2.

2. 1, 3, 6, 9 12, 15, 18, ___, ___, ___, ___, . . .

_____

3. 3, 8, 13, 18, 23, 28, 33, ___, ___, ___, ___, . . .

_____

4. 5, 12, 19, 26, 33, ___, ___, ___, ___, . . .

_____

5. 1, 4, 9, 16, 25, 36, ___, ___, ___, ___, . . .

_____

6. 50, 47, 44, 41, 38, 35, ___, ___, ___, ___, . . .

_____

7. 38, 40, 37, 39, 36, 38, 35, ___, ___, ___, ___, . . .

_____

8. 1, 4, 16, 64, ___, ___, ___, ___, . . .

_____

9. 2, 6, 18, 54, ___, ___, ___, ___, . . .

_____

10. $\frac{1}{2}$, 1, $1\frac{1}{2}$, 2, $2\frac{1}{2}$, 3, $3\frac{1}{2}$, ____, ____, ____, ____, . . .

11. 2, 1, $\frac{1}{2}$, $\frac{1}{4}$, $\frac{1}{8}$, $\frac{1}{16}$, ____, ____, ____, ____, . . .

12. 1, 2, 4, 8, 16, 32, 64, ____, ____, ____, ____, . . .

13. ____, ____, 5, 7, 11, ____, ____, ____, . . .

14. ____, ____, ____, 51, 153, 459, 1368, . . .

15. 3, 8, 13, ____, ____, 28, ____, . . .

16. 3, $\frac{1}{3}$, 4, $\frac{1}{4}$, 5, $\frac{1}{5}$, ____, ____, ____, . . .

17. 1, 2, 1, 1, 2, 1, 1, ____, ____, ____, . . .

18. 1, 1, 2, 3, 5, 8, ____, ____, . . .

19. 3, 5, 7, 9, 11, ____, . . .

20. 121, 242, 484, 968, ____, ____, . . .

Find the Pattern II

Look at the sets of numbers. Find the pattern for each set and fill in the blanks. What is the pattern? Write it on the line provided.

Example:

1. 2, 4, 8, <u>16</u>, <u>32</u>, <u>64</u>, . . .

$a, a^2, a^3, a^4, a^5, a^6$ _____

2. 2, 4, 6, 8, 2, 4, 6, 8, 2, 4, ___, ___, . . .

_____

3. 1, 3, 5, 7, 9, ___, ___, ___, . . .

_____

4. 2, 8, 14, 20, ___, ___, ___, . . .

_____

5. 3, 6, 9, 12, ___, ___, ___, ___, . . .

_____

6. 5, 10, 15, 20, ___, ___, ___, . . .

_____

7. 6, 12, 18, 24 ___, ___, ___, . . .

_____

8. 9, 12, 15, 18, ___, ___, ___, . . .

_____

9. 2, 4, 6, 2, 4, 6, 2, 4, 6, ___, ___, ___, . . .

_____

10. 1, 3, 9, 9, 3, 1, 1, ___, ___, ___, . . .

_____

11. 3, 8, 13, ____, ____, 28, ____, . . .

_____

12. $\frac{1}{2}$, 1, $1\frac{1}{2}$, 2, $2\frac{1}{2}$, 3, $3\frac{1}{2}$, ____, ____, ____, ____, . . .

_____

13. 100, 95, 90, 85, ____, ____, ____, . . .

_____

14. ____, ____, 3, 4, 5, ____, ____, ____, . . .

_____

15. ____, ____, 3, 7, 13, ____, ____, . . .

_____

16. $\frac{1}{3}$, $\frac{2}{3}$, $\frac{3}{3}$, $\frac{4}{3}$, ____, ____, ____, . . .

_____

17. $\frac{1}{3}$, $\frac{3}{6}$, $\frac{6}{9}$, $\frac{9}{12}$, ____, ____, ____, . . .

_____

18. 2, 4, 8, ____, ____, ____, . . .

_____

19. ____, ____, 30, 45, 60, . . .

_____

20. 3, 9, 27, 81, ____, 729, ____, . . .

_____

Find the Outsider I

Look at the sets of numbers. Circle the number in each set that does not belong. Give a reason for your answer.

Reason

Example:

1. ⑨ 36 74 25 48   It has only one digit.

2. 11 13 15 19 35

3. 40 6 8 12 47

4. 1 2 3 4 9

5. $\frac{1}{2}$ $\frac{1}{3}$ $\frac{1}{4}$ $\frac{2}{5}$ $\frac{1}{16}$

6. 12 36 74 25 48

7. 123 456 789 101112 954

8. 25 121 144 235 256

9. 5 11 51 10 17

10. 18 37 45 9 21

Find the Outsider II

Look at the sets of numbers. Circle the number in each set that does not belong. Give a reason for your answer.

Reason

Example:

1. 4 ⑤ 9 12 16   5 is a prime number.

2. 24 30 16 49 57 _____

3. 10 20 30 40 60 50 _____

4. 3 4 4 6 8 9 _____

5. $\frac{1}{2}$ $\frac{1}{3}$ $\frac{1}{5}$ $\frac{1}{8}$ $\frac{1}{16}$ $\frac{2}{3}$ _____

6. 0.3 0.30 0.300 0.3000 3.0 _____

7. 246 468 6,810 357 _____

8. 9 16 25 36 47 49 81 _____

9. 1 17 92 36 54 _____

10. 18 72 36 45 28 171 _____

Find the Outsider III

Look at the sets of numbers. Circle the number in each set that does not belong. Give a reason for your answer.

Reason

Example:

1. $\frac{1}{2}$ $\frac{1}{3}$ $\frac{1}{5}$ $\frac{1}{8}$ $\frac{1}{16}$ ⓔ$\frac{2}{3}$ $\frac{2}{3}$ is not a unit fraction. _____

2. $\frac{1}{2}$ $\frac{2}{3}$ $\frac{4}{6}$ $\frac{5}{7}$ $\frac{1}{8}$ _____

3. 6%   10%   13%   40%   65%   _____

4. $\frac{1}{6}$ $\frac{1}{3}$ $\frac{3}{8}$ $\frac{1}{7}$ $\frac{2}{3}$ _____

5. $\frac{1}{2}$ $\frac{2}{3}$ $\frac{5}{8}$ $\frac{5}{6}$ $\frac{8}{7}$ $\frac{7}{8}$ _____

6. $\frac{12}{24}$ $\frac{6}{12}$ $\frac{3}{6}$ $\frac{1}{3}$ $\frac{1}{2}$ _____

7. 20 $\frac{2}{10}$ $\frac{1}{5}$ 20%   2.0   _____

8. $\frac{2}{3}$ $\frac{4}{6}$ $\frac{12}{15}$ 66 $66\frac{1}{3}$ _____

9. .123%   1.23%   12.3%   123%   _____

10. $\frac{2}{5}$ $\frac{3}{4}$ $\frac{5}{8}$ $\frac{7}{8}$ $\frac{5}{6}$ _____

Find the Outsider IV

Look at the sets of numbers. Circle the number
in each set that does not belong. Give a reason for
your answer.

Reason

Example:

1. $\frac{1}{2}$ $\frac{2}{3}$ $\frac{7}{8}$ $\frac{9}{10}$ $\frac{11}{12}$ $\boxed{\frac{7}{5}}$ <u>Numerator should be 1 less than denominator.</u>

2. .63 .64 .65 .70 6.8 _____

3. 25% 50% 30% 65% 27% _____

4. $\frac{2}{3}$ $\frac{3}{3}$ $\frac{4}{3}$ $\frac{1}{3}$ $\frac{2}{5}$ $\frac{5}{3}$ _____

5. $\frac{3}{4}$ $\frac{4}{5}$ $\frac{5}{6}$ $\frac{7}{8}$ $\frac{6}{7}$ $\frac{9}{10}$ _____

6. $\frac{1}{3}$ $\frac{2}{6}$ $\frac{3}{9}$ $\frac{4}{12}$ $\frac{5}{15}$ $\frac{6}{12}$ _____

7. .30 $\frac{3}{10}$ $\frac{33}{100}$ $\frac{1}{3}$ _____

8. $\frac{3}{5}$ $\frac{6}{10}$ $\frac{9}{15}$ $\frac{12}{20}$ $\frac{16}{25}$ _____

9. .135% 1.35% 13.5% 153% _____

10. $\frac{3}{8}$ $\frac{4}{8}$ $\frac{5}{8}$ $\frac{6}{8}$ $\frac{7}{8}$ $\frac{8}{7}$ _____

Finding Patterns in Number Sets

Birds of a Feather I

Look at the sets of numbers. What do the three numbers in each set have in common? Fill in the blank.

What do they have in common?

Example:

1. 2 4 6   They can all be divided by 2.

2. 3 9 12   _____

3. 13 17 15   _____

4. 24 36 58   _____

5. 25 35 45   _____

6. 19 23 47   _____

7. 121 363 1,001 _____

8. 4 16 256   _____

9. 244 361 181   _____

10. 27 45 36   _____

Birds of a Feather II

Look at the sets of numbers. What do the three
numbers in each set have in common? Fill in the
blank.

What do they have in common?

Example:

1. 9 27 55   <u>They are all composite numbers.</u>

2. 3 6 12   _____

3. 104 68 92   _____

4. 98 175 161   _____

5. 32 112 80   _____

6. 123 468 595   _____

7. 104 560 2,560   _____

8. 53 57 58   _____

9. 162 322 483   _____

10. 17 41 437   _____

Birds of a Feather III

Look at the sets of numbers. What do the three numbers in each set have in common? Fill in the blank.

What do they have in common?

Example:

1. 4   9   25   They are all perfect squares. _____

2. 8   27   64   _____

3. $\frac{1}{2}$ $\frac{3}{8}$ $\frac{7}{16}$ _____

4. 3   9   17   _____

5. 4   40   18   _____

6. 2   8   16   _____

7. 1   19   37   _____

8. 6   12   19   _____

9. 7   11   14   _____

10. 7   13   21   _____

Birds of a Feather IV

Look at the sets of numbers. What do the three numbers in each set have in common? Fill in the blank.

What do they have in common?

Example:

1.	1	5	9	They are all odd numbers.
2.	12	45	78	
3.	23	56	89	
4.	14	25	36	
5.	47	58	69	
6.	15	26	48	
7.	24	57	68	
8.	17	28	39	
9.	51	62	95	
10.	80	69	47	

Units, Tens, Hundreds, Thousands I

Look at the numbers below. Find how many units, tens, hundreds, and thousands there are in each number. The first one has been done for you!

		Thousands	Hundreds	Tens	Units
1.	5,280	5	2	8	0
2.	3,102				
3.	57				
4.	600				
5.	5,010				
6.	1,684				
7.	523				
8.	1,010				
9.	3,303				
10.	707				

There are five thousand, two hundred eighty feet in a mile.

INCHWORM ROAD RACE

Units, Tens, Hundreds, Thousands II

Look at the numbers below. Find how many units, tens, hundreds, and thousands there are in each number. The first one has been done for you!

		Thousands	Hundreds	Tens	Units
1.	6,006	6	0	0	6
2.	5,407				
3.	1,005				
4.	3,610				
5.	2,938				
6.	3,007				
7.	4,506				
8.	3,076				
9.	4,903				
10.	6,010				

There are five thousand, two hundred eighty feet in a mile.

INCHWORM ROAD RACE

Units, Tens, Hundreds, Thousands III

Look at the problems below. Find the solution by multiplying each set and then adding them together. The first one has been done for you!

1. 3 (1,000) + 4 (100) + 5 (10) + 6 (1)   = ___3,456___

2. 3 (1,000) + 0 (100) + 5 (10) + 6 (1)   = _____

3. 3 (1,000) + 4 (100) + 0 (10) + 6 (1)   = _____

4. 3 (1,000) + 4 (100) + 5 (10) + 0 (1)   = _____

5. 0 (1,000) + 4 (100) + 5 (10) + 6 (1)   = _____

6. 1 (1,000) + 0 (100) + 0 (10) + 6 (1)   = _____

7. 1 (1,000) + 4 (100) + 5 (10) + 0 (1)   = _____

8. 3 (1,000) + 4 (100) + 0 (10) + 7 (1)   = _____

9. 1 (1,000) + 0 (100) + 6 (10) + 0 (1)   = _____

10. 0 (1,000) + 1 (100) + 0 (10) + 1 (1)   = _____

Units, Tens, Hundreds, Thousands IV

Look at the problems below. Find the solution by
multiplying each set and then adding them together.
The first one has been done for you!

1. 6 x 100   +   5 x 10   +   8 x 1   = _____658_____

2. 7 x 100   +   7 x 1   = _____

3. 9 x 100   +   5 x 10   +   5 x 1   = _____

4. 1 x 1,000 +   2 x 100 +   1 x 1   = _____

5. 11 x 1,000 +   2 x 10   +   6 x 1   = _____

6. 7 x 100   +   3 x 10   +   3 x 1   = _____

7. 3 x 1,000 +   9 x 100 +   3 x 10   +   6 x 1   = _____

8. 2 x 100   +   9 x 10   +   1 x 1   = _____

9. 7 x 1,000 +   6 x 100 +   2 x 10   +   5 x 1   = _____

10. 2 x 1,000 +   8 x 10   +   9 x 1   = _____

There are five thousand, two hundred eighty feet in a mile.

INCHWORM ROAD RACE

Rounding is the <u>KEY!</u> I

Round off each number to the nearest target number. For example, is 96 closer to 90 or 100? (The first one has been done for you!)

	Number	Target number	Rounds to:
1.	9	Nearest 10	_10_
2.	5	Nearest 10	_____
3.	4	Nearest 10	_____
4.	38	Nearest 10	_____
5.	27	Nearest 10	_____
6.	64	Nearest 100	_____
7.	152	Nearest 100	_____
8.	321	Nearest 100	_____
9.	763	Nearest 100	_____
10.	501	Nearest 100	_____
11.	$\frac{1}{2}$	Nearest whole number	_____
12.	$\frac{2}{3}$	Nearest whole number	_____
13.	$\frac{5}{8}$	Nearest whole number	_____
14.	$\frac{3}{7}$	Nearest whole number	_____
15.	$\frac{2}{5}$	Nearest whole number	_____

Rounding is the <u>KEY!</u> II

Look at the numbers below. Round off each
number to the nearest target number. The
first one has been done for you!

	Number	Target number	Rounds to:
1.	107	Nearest 100	<u>100</u>
2.	259	Nearest 100	_____
3.	249	Nearest 100	_____
4.	681	Nearest 100	_____
5.	750	Nearest 100	_____
6.	3,527	Nearest 1,000	_____
7.	4,321	Nearest 1,000	_____
8.	8,664	Nearest 1,000	_____
9.	4,032	Nearest 1,000	_____
10.	3,102	Nearest 1,000	_____
11.	$\frac{4}{9}$	Nearest whole number	_____
12.	$\frac{13}{18}$	Nearest whole number	_____
13.	$6\frac{2}{3}$	Nearest whole number	_____
14.	$5\frac{3}{8}$	Nearest whole number	_____
15.	$7\frac{1}{2}$	Nearest whole number	_____

Rounding is the <u>KEY!</u> III

Look at the numbers below. Round off each number to the nearest target number. The first one has been done for you!

	Number	Target number	Rounds to:
1.	38	Nearest 10	_40_
2.	46	Nearest 10	_____
3.	132	Nearest 10	_____
4.	257	Nearest 10	_____
5.	671	Nearest 10	_____
6.	235	Nearest 100	_____
7.	356	Nearest 100	_____
8.	795	Nearest 100	_____
9.	846	Nearest 100	_____
10.	960	Nearest 100	_____
11.	$\frac{2}{3}$	Nearest whole number	_____
12.	$\frac{5}{6}$	Nearest whole number	_____
13.	$\frac{5}{8}$	Nearest whole number	_____
14.	$\frac{1}{3}$	Nearest whole number	_____
15.	$7\frac{2}{3}$	Nearest whole number	_____

Rounding is the __KEY!__ IV

Look at the numbers below. Round off each number to the nearest target number. The first one has been done for you!

	Number	Target number	Rounds to:
1.	158	Nearest 100	_200_
2.	204	Nearest 100	_____
3.	333	Nearest 100	_____
4.	531	Nearest 100	_____
5.	699	Nearest 100	_____
6.	3,647	Nearest 1,000	_____
7.	4,241	Nearest 1,000	_____
8.	6,840	Nearest 1,000	_____
9.	2,840	Nearest 1,000	_____
10.	1,409	Nearest 1,000	_____
11.	$\frac{4}{9}$	Nearest whole number	_____
12.	$\frac{13}{18}$	Nearest whole number	_____
13.	$6\frac{2}{3}$	Nearest whole number	_____
14.	$5\frac{3}{8}$	Nearest whole number	_____
15.	$7\frac{1}{3}$	Nearest whole number	_____

Estimate the Answer I

Use your skills of rounding off to estimate the following answers. What did you think about? (The first one has been done for you!)

ahh . . . about 50 plus about 60 is about 110!

Problem:	What I thought about:	My estimate:
1. 46 + 59	50 + 60	110
2. 38 + 120		
3. 59 + 72		
4. 68 + 95		
5. 480 − 49		
6. 699 − 428		
7. 384 − 249		
8. 499 − 348		
9. 63 x 24		
10. 38 x 42		
11. 93 x 42		
12. 127 x 156		
13. 346 ÷ 23		
14. 498 ÷ 34		
15. 864 ÷ 49		
16. 1,374 ÷ 58		

Estimate the Answer II

Use your skills of rounding off to estimate the following answers. What did you think about? (The first one has been done for you!)

ahh . . . about 50 plus about 60 is about 110!

	Problem:	What I thought about:	My estimate:
1.	63 + 99	60 + 100	160
2.	78 + 42		
3.	38 + 42		
4.	96 + 69		
5.	561 − 38		
6.	752 − 321		
7.	126 − 49		
8.	364 − 42		
9.	52 x 38		
10.	36 x 44		
11.	28 x 32		
12.	125 x 125		
13.	38 ÷ 6		
14.	59 ÷ 4		
15.	123 ÷ 26		
16.	1,426 ÷ 32		

Estimate the Answer III

Use your skills of rounding off to estimate the following answers. What did you think about? (The first one has been done for you!)

ahh . . . about 50 plus about 60 is about 110!

	Problem:	What I thought about:	My estimate:
1.	37 + 68	*40 + 70*	*110*
2.	42 + 71		
3.	125 + 231		
4.	263 + 347		
5.	98 − 42		
6.	75 − 39		
7.	136 − 43		
8.	247 − 152		
9.	17 x 23		
10.	32 x 56		
11.	84 x 62		
12.	127 x 38		
13.	47 ÷ 7		
14.	93 ÷ 8		
15.	127 ÷ 13		
16.	2,347 ÷ 157		

Estimate the Answer IV

Use your skills of rounding off to estimate the following answers. What did you think about? (The first one has been done for you!)

ahh . . . about 50 plus about 60 is about 110!

Problem:	What I thought about:	My estimate:
1. 16 ÷ 5	20 ÷ 5	4
2. 19 ÷ 4		
3. 28 ÷ 6		
4. 134 ÷ 31		
5. 94 + 38		
6. 163 + 47		
7. 58 + 37		
8. 433 + 129		
9. 8 x 13		
10. 7 x 21		
11. 15 x 16		
12. 123 x 36		
13. 129 ÷ 64		
14. 129 ÷ 4		
15. 1,321 ÷ 6		
16. 1,460 ÷ 70		

Find the Ones Digit I

		"Ones" Digit			"Ones" Digit
Example:					
1.	17 + 19	6	**14.**	94 x 37	____
2.	38 + 47	____	**15.**	132 x 68	____
3.	136 + 249	____	**16.**	68 x 4	____
4.	378 + 692	____	**17.**	132 x 8	____
5.	1,547 + 2,394	____	**18.**	6,429 x 3	____
6.	98 – 45	____	**19.**	4,725 x 5	____
7.	116 – 42	____	**20.**	9,643 x 41	____
8.	336 – 137	____	**21.**	1,246 + 3,964	____
9.	578 – 426	____	**22.**	3,729 – 2,435	____
10.	416 – 147	____	**23.**	998 – 642	____
11.	38 x 6	____	**24.**	3,846 x 42	____
12.	42 x 38	____	**25.**	39 x 86	____
13.	64 x 85	____			

Seventeen Fishburgers, no mayo.

Nineteen Fishburgers, travelin'.

OK, Seven plus nine equals Six...

Find the Ones Digit II

		"Ones" Digit			**"Ones" Digit**
Example:					
1.	9 + 6	5	**14.**	63 x 9	_____
2.	18 + 21	_____	**15.**	73 x 8	_____
3.	36 + 77	_____	**16.**	86 x 4	_____
4.	44 + 98	_____	**17.**	93 x 7	_____
5.	65 + 68	_____	**18.**	126 x 42	_____
6.	106 – 42	_____	**19.**	38 x 68	_____
7.	98 – 39	_____	**20.**	44 x 44	_____
8.	136 – 48	_____	**21.**	56 x 94	_____
9.	562 – 134	_____	**22.**	3,211 – 1,207	_____
10.	168 – 49	_____	**23.**	336 – 48	_____
11.	28 x 9	_____	**24.**	48 x 64	_____
12.	48 x 6	_____	**25.**	44 x 96	_____
13.	58 x 7	_____			

Seventeen Fishburgers, no mayo.

Nineteen Fishburgers, travelin'.

OK, seven plus nine equals six...

Estimate and Find the "Ones" Digit I

Use your skills of rounding off to estimate the following answers. What did you think about? (The first one has been done for you!)

	Problem:	**Estimate:**	**"Ones" digit:**
1.	47 + 38	50 + 40 = 90	5
2.	65 + 127		
3.	127 + 364		
4.	948 + 347		
5.	1,347 + 2,398		
6.	3,969 − 2,396		
7.	4,738 − 3,444		
8.	9,364 − 4,235		
9.	8,464 − 3,942		
10.	10,376 − 9,342		
11.	16 x 17		
12.	21 x 19		
13.	38 x 64		
14.	59 x 88		
15.	184 x 236		

16. 1,367 x 248   _____   _____

17. 171 x 42   _____   _____

18. 711 x 24   _____   _____

19. 117 x 32   _____   _____

20. 141 x 64   _____   _____

21. 88 x 88   _____   _____

22. 49 + 86   _____   _____

23. 79 x 147   _____   _____

24. 36 x 47   _____   _____

25. 81 x 40   _____   _____

26. 19 + 17   _____   _____

27. 48 − 16   _____   _____

28. 25 x 26   _____   _____

29. 38 x 16   _____   _____

30. 14 x 14   _____   _____

31. 36 − 16   _____   _____

32. 158 ÷ 2   _____   _____

33. 1,233 ÷ 3   _____   _____

34. 1,616 ÷ 4   _____   _____

35. 5,608 ÷ 8   _____   _____

Estimate and Find the "Ones" Digit II

Use your skills of rounding off to estimate the
following answers. What did you think about?
(The first one has been done for you!)

	Problem:	Estimate:	"Ones" digit:
1.	17 + 19	20 + 20 = 40	6
2.	24 + 36		
3.	27 + 54		
4.	138 + 64		
5.	238 + 346		
6.	4,912 − 3,046		
7.	3,536 − 2,934		
8.	6,934 − 4,932		
9.	7,938 − 6,348		
10.	10,364 − 4,962		
11.	15 x 16		
12.	23 x 47		
13.	38 x 26		
14.	57 x 64		
15.	138 x 64		

16. 1,384 x 52   _____   _____

17. 176 x 38   _____   _____

18. 73 x 48   _____   _____

19. 163 x 21   _____   _____

20. 362 x 16   _____   _____

21. 93 x 56   _____   _____

22. 45 x 64   _____   _____

23. 38 + 136   _____   _____

24. 64 + 39   _____   _____

25. 136 + 38   _____   _____

26. 168 ÷ 8   _____   _____

27. 240 ÷ 7   _____   _____

28. 363 ÷ 4   _____   _____

29. 439 ÷ 5   _____   _____

30. 1,603 ÷ 10   _____   _____

31. 432 ÷ 5   _____   _____

32. 630 ÷ 7   _____   _____

33. 117 ÷ 9   _____   _____

34. 333 ÷ 6   _____   _____

35. 4,032 ÷ 8   _____   _____

Compatible Numbers 1

Look at the problems below. Use your skills of rounding off to estimate the answer in two different ways. What did you think about? (The first one has been done for you!)

Problem:	First solution:	My estimate:
1. 564 – 119 •	119 + 1 = 120	430
•	564 – 120 = 444	
•	444 + 1 = 445	

Problem:	Second solution:	My estimate:
564 – 119 •	520 – 120 = 400	400
•	564 – 520 = 44	
•	400 + 44 + 1 = 445	

Problem:	First solution:	My estimate:
2. 137 + 259 •	_____	_____
•	_____	
•	_____	

Problem:	Second solution:	My estimate:
137 + 259 •	_____	_____
•	_____	
•	_____	

Problem:	**First solution:**	**My estimate:**
3. 68 + 75	• _____ • _____ • _____	_____

Problem:	**Second solution:**	**My estimate:**
68 + 75	• _____ • _____ • _____	_____

Problem:	**First solution:**	**My estimate:**
4. 38 + 42	• _____ • _____ • _____	_____

Problem:	**Second solution:**	**My estimate:**
38 + 42	• _____ • _____ • _____	_____

Compatible Numbers II

Look at the problems below. Use your skills of rounding off to estimate the answer in two different ways. What did you think about? (The first one has been done for you!)

Problem:	**First solution:**	**My estimate:**
1. 68 x 39	• _68 (40 − 1)_	_70 x 40 = 2800_
	• _2720 − 68 =_	
	• _2652_	

Problem:	**Second solution:**	**My estimate:**
68 x 39	• _39 (60 + 8)_	_2800_
	• _2340 = 320 x 8 =_	
	• _2660 − 8 = 2652_	

Problem:	**First solution:**	**My estimate:**
2. 138 ÷ 6	• _____	_____
	• _____	
	• _____	

Problem:	**Second solution:**	**My estimate:**
138 ÷ 6	• _____	_____
	• _____	
	• _____	

Problem:	First solution:	My estimate:
3. $249 \div 7$	• _____	_____
	• _____	
	• _____	

Problem:	Second solution:	My estimate:
$249 \div 7$	• _____	_____
	• _____	
	• _____	

Problem:	First solution:	My estimate:
4. 130×64	• _____	_____
	• _____	
	• _____	

Problem:	Second solution:	My estimate:
130×64	• _____	_____
	• _____	
	• _____	

Compatible Numbers III

Look at the problems below. Use your skills of rounding off to make compatible numbers. Then try to answer the problems without writing them. What did you think about? (The first one has been done for you!)

	Problem:	**Solution:**	**Explanation:**
1.	89 + 47	136	90 + 50 = 140 and 140 − 4 = 136
2.	135 + 437	_____	
3.	407 − 135	_____	
4.	316 − 147	_____	
5.	35 x 28	_____	
6.	52 x 48	_____	
7.	216 ÷ 6	_____	
8.	440 ÷ 8	_____	
9.	240 ÷ 12	_____	
10.	156 ÷ 6	_____	

Compatible Numbers IV

Look at the problems below. Use your skills of rounding off to make compatible numbers. Then try to answer the problems without writing them. What did you think about? (The first one has been done for you!)

	Problem:	**Solution:**	**Explanation:**
1.	38 + 62	100	40 + 60 = 100 and 100 + 2 - 2 = 100
2.	49 + 71	_____	_____
3.	638 − 132	_____	_____
4.	321 − 108	_____	_____
5.	25 x 42	_____	_____
6.	50 x 21	_____	_____
7.	68 ÷ 8	_____	_____
8.	124 ÷ 6	_____	_____
9.	368 ÷ 8	_____	_____
10.	168 ÷ 8	_____	_____

Now That Makes Sense! I

Look at the following statements. Check "Yes" if the statement makes sense. Check "No" if the statement does not make sense. Write your reason on the line below.

		Yes	No
Example:			
1.	There are 100 quarters in $25.00.	✔	____
	There are four quarters in a dollar. 4 x 25 = 100.		
2.	A person could throw a baseball 18,000 inches.	____	____
3.	A person could be 200 inches tall.	____	____
4.	A car could travel 1 mile per minute.	____	____
5.	A basketball could weigh 40 pounds.	____	____
6.	There are 1,000 pennies in $10.00.	____	____
7.	An airplane could fly at an altitude of 40,000 feet.	____	____
8.	A person can give 110% of their energy to a game.	____	____
9.	The difference between two numbers is greater than either number.	____	____
10.	A coin could be flipped and land on its edge.	____	____

		Yes	No

11. Three numbers added together can total less than 1. _____ _____

_____

12. You could collect 1 million bottle caps. _____ _____

_____

13. There are 100 dimes in $10.00. _____ _____

_____

14. A number could be divisible by 2, 3, 4, and 5. _____ _____

_____

15. Two fractions added together could give you less
than either number. _____ _____

_____

16. A person could hit a golf ball 2,000 feet. _____ _____

_____

17. A person could eat $\frac{3}{2}$ of a pizza. _____ _____

_____

18. One half plus one half equals one fourth. _____ _____

_____

19. There are 6 halves in 3 pizzas. _____ _____

_____

20. A person could ride a bicycle at a speed of 40 mph. _____ _____

_____

21. A person could eat 100 pounds of ice cream. _____ _____

_____

22. A student could get 100% on every test. _____ _____

_____

23. A man could throw a football 2,900 inches. _____ _____

_____

24. The area and perimeter of a figure could be equal. _____ _____

_____

25. A person could live 10 decades. _____ _____

_____

Now That Makes Sense! II

Look at the following estimates. Check "Yes" if the statement makes sense. Check "No" if the statement does not make sense.

	Example:		**Yes**	**No**
1.	11 x 12 is approximately:	100	✔	____
2.	14 x 16 is approximately:	30	____	____
3.	19 x 21 is approximately:	40	____	____
4.	48 x 52 is approximately:	2,500	____	____
5.	64 x 38 is approximately:	100	____	____
6.	96 x 48 is approximately:	30	____	____
7.	63 – 27 is approximately:	40	____	____
8.	1,148 – 211 is approximately:	900	____	____
9.	196 – 48 is approximately:	15	____	____
10.	361 ÷ 5 is approximately:	7	____	____
11.	147 ÷ 9 is approximately:	6	____	____
12.	634 ÷ 11 is approximately:	6	____	____

13. $\frac{1}{2} + \frac{1}{3}$ is approximately: $\frac{2}{5}$ _____ _____

14. $\frac{1}{8} + \frac{5}{6}$ is approximately: **1** _____ _____

15. $1\frac{1}{2} + 2\frac{1}{3}$ is approximately: $\mathbf{3\frac{1}{2}}$ _____ _____

16. $6\frac{2}{3} + 5\frac{1}{2}$ is approximately: **12** _____ _____

17. $\frac{3}{4} - \frac{1}{2}$ is approximately: $\frac{2}{5}$ _____ _____

18. $\frac{6}{7} - \frac{1}{3}$ is approximately: $\frac{1}{2}$ _____ _____

19. $\frac{9}{10} - \frac{2}{3}$ is approximately: $\frac{1}{2}$ _____ _____

20. $3\frac{1}{2} - 2\frac{1}{3}$ is approximately: **1** _____ _____

21. $\frac{1}{2} + \frac{1}{8}$ is approximately: $\frac{1}{16}$ _____ _____

22. $\frac{3}{4} - \frac{2}{3}$ is approximately: **1** _____ _____

23. $3\frac{1}{2} + 4\frac{1}{2}$ is approximately: **7** _____ _____

24. $8 \times \frac{1}{2}$ is approximately: **16** _____ _____

25. $\frac{1}{2} \div \frac{1}{4}$ is approximately: **2** _____ _____

26. $\frac{2}{3} \div \frac{1}{3}$ is approximately: **2** _____ _____

27. $\frac{2}{5} \div \frac{1}{3}$ is approximately: **1** _____ _____

28. $\frac{7}{8} \div \frac{3}{8}$ is approximately: **2** _____ _____

29. $\frac{2}{7} \div \frac{3}{7}$ is approximately: $\frac{5}{14}$ _____ _____

Now That Makes Sense! III

Look at the following exercises. These answers were
given by students. Check "Yes" if the answer makes
sense. Check "No" if the answer does not make sense.
Explain your reason.

		Yes	No	Explanation

Example:

1.
$$\begin{array}{r} 365 \\ + 296 \\ \hline 663 \end{array}$$

✔ ____   Less than four hundred plus about three hundred equals less than seven hundred.

2.
$$\begin{array}{r} 456 \\ + 978 \\ \hline 1,324 \end{array}$$

____ ____

3.
$$\begin{array}{r} 567 \\ + 392 \\ \hline 859 \end{array}$$

____ ____

4.
$$\begin{array}{r} 648 \\ + 429 \\ \hline 1,067 \end{array}$$

____ ____

5.
$$\begin{array}{r} 998 \\ + 647 \\ \hline 1,635 \end{array}$$

____ ____

Now That Makes Sense! IV

Look at the following exercises. These answers were given by students. Check "Yes" if the answer makes sense. Check "No" if the answer does not makes sense. Explain your reason.

		Yes	No	Explanation

1. 373
+ 296
573 — 569

2. 6,432
+ 948
6,370

3. 7,032
+ 3,064
10,096

4. 3,232
+ 6,392
9,624

5. 40,378
+ 6,472
46,740

Now That Makes Sense! V

Look at the following exercises. These answers were
given by students. Check "Yes" if the answer makes
sense. Check "No" if the answer does not make sense.
Explain your reason.

	Yes	No	Explanation

1.   638
  − 240
   878

2.   1,039
  − 642
   417

3.   938
  − 425
   513

4.   638
  − 421
   217

5.   987
  − 964
    23

Now That Makes Sense! VI

Look at the following exercises. These answers were
given by students. Check "Yes" if the answer makes
sense. Check "No" if the answer does not make sense.
Explain your reason.

	Yes	No	Explanation

1. 1,325
− 832
593 _____ _____ _____

2. 2,346
− 2,001
345 _____ _____ _____

3. 3,296
− 2,407
6,703 _____ _____ _____

4. 6,429
− 3,216
3,213 _____ _____ _____

5. 7,324
− 4,231
3,113 _____ _____ _____

Reasonable or Unreasonable? 1

Look at the following exercises. If the answer is close to the correct answer, check "Reasonable." If the answer is really far off from the correct answer, check "Unreasonable." Explain your answer in the space provided.

		Reasonable	**Unreasonable**	**Explanation**

1.
$$\begin{array}{r} 32 \\ \times\ 46 \\ \hline 78 \end{array}$$
Reasonable _____ Unreasonable _____

2.
$$\begin{array}{r} 28 \\ \times\ 45 \\ \hline 1{,}260 \end{array}$$
Reasonable _____ Unreasonable _____

3.
$$\begin{array}{r} 52 \\ \times\ 31 \\ \hline 1{,}021 \end{array}$$
Reasonable _____ Unreasonable _____

4.
$$\begin{array}{r} 126 \\ \times\ \ 4 \\ \hline 500 \end{array}$$
Reasonable _____ Unreasonable _____

5.
$$\begin{array}{r} 62 \\ \times\ 14 \\ \hline 868 \end{array}$$
Reasonable _____ Unreasonable _____

Reasonable or Unreasonable? II

Look at the following exercises. If the answer is close to the correct answer, check "Reasonable." If the answer is really far off from the correct answer, check "Unreasonable." Explain your answer in the space provided.

	Reasonable	Unreasonable	Explanation

1. 19
x 23
437
____ ____ _____
_____
_____

2. 42
x 59
2,038
____ ____ _____
_____
_____

3. 75
x 25
100
____ ____ _____
_____
_____

4. 302
x 64
1,608
____ ____ _____
_____
_____

5. 950
÷ 25
38
____ ____ _____
_____
_____

Reasonable or Unreasonable? III

Look at the following exercises. These answers were given by students. Check "Yes" if the answer makes sense. Check "No" if the answer does not make sense. Explain your reason.

	Yes	No

1. $1,012 \div 23 = 50$

Explanation:

_____

2. $2,665 \div 41 = 40$

Explanation:

_____

3. $136 \div 7 = 112$

Explanation:

_____

4. $224 \div 14 = 160$

Explanation:

_____

5. $3,276 \div 32 = 12.37$

Explanation:

_____

Reasonable or Unreasonable? IV

Look at the following exercises. These answers were given by students. Check "Yes" if the answer makes sense. Check "No" if the answer does not make sense. Explain your reason.

	Yes	No

1. 30% of 30 is 10.

Explanation:

_____

2. 100% of 17 is 17.

Explanation:

_____

3. 12 is 200% of 3.

Explanation:

_____

4. 8 is 40% of 20.

Explanation:

_____

5. 150% of 6 is 12.

Explanation:

_____

Reasonable or Unreasonable? V

Look at the following exercises. These answers were given by students. Check "Yes" if the answer makes sense. Check "No" if the answer does not make sense. Explain your reason.

	Yes	No

1. $\frac{1}{2} + \frac{1}{3} = \frac{1}{5}$ _____   _____

   Explanation:

   _____

2. $\frac{2}{3} \times \frac{1}{6} = \frac{1}{2}$ _____   _____

   Explanation:

   _____

3. $\frac{2}{3} \times \frac{1}{2} = \frac{3}{4}$ _____   _____

   Explanation:

   _____

4. $\frac{1}{2} \div \frac{1}{4} = 8$ _____   _____

   Explanation:

   _____

5. $\frac{2}{3} \div \frac{1}{6} = \frac{4}{13}$ _____   _____

   Explanation:

   _____

Reasonable or Unreasonable? VI

Look at the following exercises. These answers were given by students. Check "Yes" if the answer makes sense. Check "No" if the answer does not make sense. Explain your reason.

	Yes	No

1. $\frac{4}{7} \times \frac{1}{2} = \frac{4}{14}$ _____   _____

Explanation:

_____

2. $\frac{2}{5} + \frac{3}{8} = \frac{6}{40}$ _____   _____

Explanation:

_____

3. $\frac{6}{7} + \frac{3}{4} = \frac{9}{28}$ _____   _____

Explanation:

_____

4. $\frac{2}{3} + \frac{3}{8} = \frac{25}{24}$ _____   _____

Explanation:

_____

5. $\frac{1}{2} \div \frac{1}{3} = \frac{1}{6}$ _____   _____

Explanation:

_____

Adding Odds and Evens

Solve the following problems and look for patterns.

1. 4 + 2 = _____

2. 6 + 8 = _____

3. 8 + 18 = _____

4. 12 + 14 = _____

5. 2 + 8 = _____

6. 4 + 8 = _____

7. 6 + 6 = _____

8. 14 + 16 = _____

9. 3 + 5 = _____

10. 5 + 7 = _____

11. 7 + 9 = _____

12. 9 + 11 = _____

13. 13 + 15 = _____

14. 17 + 19 = _____

15. 5 + 9 = _____

16. 9 + 15 = _____

17. 3 + 2 = _____

18. 5 + 6 = _____

19. 7 + 12 = _____

20. 9 + 4 = _____

21. 11 + 8 = _____

22. 13 + 8 = _____

23. 15 + 14 = _____

24. 17 + 6 = _____

What kind of numbers were in problems 1–8? _____

What kind of numbers were in problems 9–16? _____

What kind of numbers were in problems 17–24? _____

What always happens when you add two **even** numbers?

_____

What always happens when you add two **odd** numbers?

_____

What always happens when you add **an even and an odd** number?

_____

Multiplying Odds and Evens

Solve the following problems and look for patterns.

1. 2 x 4 = _____

2. 2 x 6 = _____

3. 4 x 6 = _____

4. 4 x 8 = _____

5. 6 x 8 = _____

6. 8 x 12 = _____

7. 10 x 12 = _____

8. 10 x 10 = _____

9. 3 x 3 = _____

10. 3 x 5 = _____

11. 5 x 7 = _____

12. 7 x 9 = _____

13. 9 x 5 = _____

14. 9 x 3 = _____

15. 7 x 11 = _____

16. 7 x 13 = _____

17. 2 x 5 = _____

18. 4 x 7 = _____

19. 6 x 5 = _____

20. 8 x 7 = _____

21. 4 x 9 = _____

22. 2 x 7 = _____

23. 6 x 9 = _____

24. 12 x 11 = _____

What kind of numbers were in problems 1–8? _____

What kind of numbers were in problems 9–16? _____

What kind of numbers were in problems 17–24? _____

What always happens when you multiply two **even** numbers?

_____

What always happens when you multiply two **odd** numbers?

_____

What always happens when you multiply **an even and an odd** number?

_____

Subtracting Odds and Evens

Solve the following problems and look for patterns.

1. 8 – 2 = _____

2. 6 – 4 = _____

3. 10 – 6 = _____

4. 12 – 8 = _____

5. 14 – 6 = _____

6. 16 – 4 = _____

7. 18 – 10 = _____

8. 20 – 12 = _____

9. 19 – 6 = _____

10. 13 – 8 = _____

11. 15 – 4 = _____

12. 13 – 2 = _____

13. 11 – 4 = _____

14. 9 – 8 = _____

15. 7 – 6 = _____

16. 21 – 14 = _____

17. 20 – 7 = _____

18. 18 – 5 = _____

19. 14 – 7 = _____

20. 12 – 5 = _____

21. 10 – 5 = _____

22. 6 – 5 = _____

23. 16 – 9 = _____

24. 8 – 3 = _____

What kind of numbers were in problems 1–8? _____

What kind of numbers were in problems 9–16? _____

What kind of numbers were in problems 17–24? _____

What always happens when you subtract an **even** number from another **even** number?

_____

What always happens when you subtract an **odd** number from another **odd** number?

_____

What always happens when you subtract an **even** number from an **odd** number?

_____

Dividing Odds and Evens

Solve the following problems and look for patterns.

1. $6\overline{)18}$ **6.** $6\overline{)17}$

2. $4\overline{)24}$ **7.** $4\overline{)19}$

3. $6\overline{)36}$ **8.** $8\overline{)37}$

4. $8\overline{)38}$ **9.** $4\overline{)47}$

5. $4\overline{)42}$ **10.** $6\overline{)51}$

What did you notice?

_____

_____

_____

Break It Down, Build It Up! I

Solve the following problems in two different ways.

Example:

36 + 45   =   35 + 1 + 35 + 10 = 35 + 35 + 1 + 10 = 70 + 11 = 81

         =   40 − 4 + 40 + 5 = 80 + 1 = 81

1.   29 + 36   = _____

              = _____

2.   45 + 39   = _____

              = _____

3.   56 + 41   = _____

              = _____

4.   95 + 63   = _____

              = _____

5.   78 + 63   = _____

              = _____

6. 79 + 59 = _____

= _____

7. 65 + 61 = _____

= _____

8. 89 + 79 = _____

= _____

9. 123 + 47 = _____

= _____

10. 127 + 58 = _____

= _____

11. 238 + 69 = _____

= _____

12. 1,246 + 2,938 = _____

= _____

13. 639 + 999 = _____

= _____

14. 234 + 567 = _____

= _____

15. 345 + 786 = _____

= _____

Break It Down, Build It Up! II

Solve the following problems using the distributive property in two different ways.

Example:

7 (17) = $\underline{\quad 7(10 + 7) = 70 + 49 = 119 \quad}$

= $\underline{\quad 7(6 + 11) = 42 + 77 = 119 \quad}$

1. 8 (25) = $\underline{\qquad\qquad\qquad\qquad\qquad}$

= $\underline{\qquad\qquad\qquad\qquad\qquad}$

2. 6 (48) = $\underline{\qquad\qquad\qquad\qquad\qquad}$

= $\underline{\qquad\qquad\qquad\qquad\qquad}$

3. 9 (11) = $\underline{\qquad\qquad\qquad\qquad\qquad}$

= $\underline{\qquad\qquad\qquad\qquad\qquad}$

4. 7 (16) = $\underline{\qquad\qquad\qquad\qquad\qquad}$

= $\underline{\qquad\qquad\qquad\qquad\qquad}$

5. 10 (14) = $\underline{\qquad\qquad\qquad\qquad\qquad}$

= $\underline{\qquad\qquad\qquad\qquad\qquad}$

6. 10 (65) = _____

= _____

7. 12 (15) = _____

= _____

8. 15 (25) = _____

= _____

9. 3 (64) = _____

= _____

10. 5 (75) = _____

= _____

11. 5 (92) = _____

= _____

12. 9 (91) = _____

= _____

13. 8 (64) = _____

= _____

14. 9 (72) = _____

= _____

15. 13 (13) = _____

= _____

Break It Down, Build It Up! III

Solve the following problems using the distributive property in two different ways.

Example:

6 (31) = _6 (30 + 1) = 180 + 6 = 186_

= _6 (20 + 11) = 120 + 66 = 186_

1. 4 (24) = _____

= _____

2. 3 (27) = _____

= _____

3. 4 (28) = _____

= _____

4. 5 (26) = _____

= _____

5. 11 (14) = _____

= _____

6. 12 (13) = _____

         = _____

7. 6 (19) = _____

         = _____

8. 12 (25) = _____

         = _____

9. 9 (16) = _____

         = _____

10. 8 (17) = _____

          = _____

11. 5 (25) = _____

          = _____

12. 6 (38) = _____

          = _____

13. 9 (103) = _____

           = _____

14. 9 (63) = _____

          = _____

15. 12 (47) = _____

           = _____

Put Together, Take Apart I

Put together combinations in two different ways.

Example:

$4 (27)$ = $\underline{\qquad 4(25) + 4 \times 2 = 100 + 8 = 108 \qquad}$

= $\underline{\qquad 4(20 + 7) = 80 + 28 = 108 \qquad}$

1. $5 (25)$ = $\underline{\qquad\qquad\qquad\qquad\qquad\qquad\qquad\qquad}$

= $\underline{\qquad\qquad\qquad\qquad\qquad\qquad\qquad\qquad}$

2. $6 (14)$ = $\underline{\qquad\qquad\qquad\qquad\qquad\qquad\qquad\qquad}$

= $\underline{\qquad\qquad\qquad\qquad\qquad\qquad\qquad\qquad}$

3. $7 (22)$ = $\underline{\qquad\qquad\qquad\qquad\qquad\qquad\qquad\qquad}$

= $\underline{\qquad\qquad\qquad\qquad\qquad\qquad\qquad\qquad}$

4. $8 (31)$ = $\underline{\qquad\qquad\qquad\qquad\qquad\qquad\qquad\qquad}$

= $\underline{\qquad\qquad\qquad\qquad\qquad\qquad\qquad\qquad}$

5. $9 (27)$ = $\underline{\qquad\qquad\qquad\qquad\qquad\qquad\qquad\qquad}$

= $\underline{\qquad\qquad\qquad\qquad\qquad\qquad\qquad\qquad}$

6. 4 (96) = _____

= _____

7. 5 (32) = _____

= _____

8. 7 (28) = _____

= _____

9. 6 (43) = _____

= _____

10. 7 (39) = _____

= _____

11. 8 (62) = _____

= _____

12. 7 (49) = _____

= _____

13. 9 (96) = _____

= _____

14. 12 (13) = _____

= _____

15. 16 (18) = _____

= _____

Put Together, Take Apart II

Put together combinations in two different ways.

Example:

6 (38)   =   _60 (30 + 8) = 6 × 30 + 6 × 8 = 180 + 48 = 228_

         =   _6 (40 – 2) = 240 – 12 = 228_

1.   7 (124)   =   _____

         =   _____

2.   12 (25)   =   _____

         =   _____

3.   25 (63)   =   _____

         =   _____

4.   37 (41)   =   _____

         =   _____

5.   64 (29)   =   _____

         =   _____

6. 25 (63) = _____

= _____

7. 64 (29) = _____

= _____

8. 39 (37) = _____

= _____

9. 16 (16) = _____

= _____

10. 19 (21) = _____

= _____

11. 32 (34) = _____

= _____

12. 126 (51) = _____

= _____

13. 250 (62) = _____

= _____

14. 314 (48) = _____

= _____

15. 16 (18) = _____

= _____

Squares and Patterns 1

Complete problems 1–10 and look for a pattern.

Example:

1. 9 x 9 = _81_

8 x 10 = _80_

2. 7 x 7 = _____

6 x 8 = _____

3. 5 x 5 = _____

4 x 6 = _____

4. 10 x 10 = _____

9 x 11 = _____

5. 11 x 11 = _____

10 x 12 = _____

6. 4 x 4 = _____

3 x 5 = _____

7. 6 x 6 = _____

5 x 7 = _____

8. 13 x 13 = _____

12 x 14 = _____

9. 17 x 17 = _____

16 x 18 = _____

10. 20 x 20 = _____

19 x 21 = _____

Now complete problems 11–20 using the pattern.
What is your discovery? Write it at the end.

11. $30 \times 30 = 900$

$29 \times 31 =$ _____

12. $40 \times 40 = 1{,}600$

$39 \times 41 =$ _____

13. $50 \times 50 = 2{,}500$

$49 \times 51 =$ _____

14. $60 \times 60 = 3{,}600$

$59 \times 61 =$ _____

15. $70 \times 70 = 4{,}900$

$69 \times 71 =$ _____

16. $18 \times 18 = 324$

$17 \times 19 =$ _____

17. $23 \times 23 = 528$

$22 \times 24 =$ _____

18. $36 \times 36 = 1{,}296$

$35 \times 37 =$ _____

19. $99 \times 99 = 9{,}801$

$98 \times 100 =$ _____

20. $102 \times 102 = 10{,}404$

$101 \times 103 =$ _____

Your Discovery: _____

_____

_____

Squares and Patterns II

Complete problems 1–10 and look for a pattern.

Example:

1. 4 x 4 = _16_

2 x 6 = _12_

2. 5 x 5 = _____

3 x 7 = _____

3. 6 x 6 = _____

4 x 8 = _____

4. 7 x 7 = _____

5 x 9 = _____

5. 8 x 8 = _____

6 x 10 = _____

6. 9 x 9 = _____

7 x 11 = _____

7. 10 x 10 = _____

8 x 12 = _____

8. 11 x 11 = _____

9 x 13 = _____

9. 12 x 12 = _____

10 x 14 = _____

10. 15 x 15 = _____

13 x 17 = _____

Now complete problems 11–20 using the pattern.
What is your discovery? Write it at the end.

11. 20 x 20 = _____

18 x 22 = _____

12. 30 x 30 = _____

28 x 32 = _____

13. 40 x 40 = _____

38 x 42 = _____

14. 50 x 50 = _____

48 x 52 = _____

15. 60 x 60 = _____

58 x 62 = _____

16. 70 x 70 = _____

68 x 72 = _____

17. 80 x 80 = _____

78 x 82 = _____

18. 90 x 90 = _____

88 x 92 = _____

19. 100 x 100 = _____

98 x 102 = _____

20. 107 x 107 = _11,449_

105 x 109 = _____

Your Discovery: _____

_____

_____

Squares and Patterns III

Complete problems 1–10 and look for a pattern.

Example:

1. 9 x 9 = _81_

   6 x 12 = _72_

2. 8 x 8 = _____

   5 x 11 = _____

3. 7 x 7 = _____

   4 x 10 = _____

4. 6 x 6 = _____

   3 x 9 = _____

5. 5 x 5 = _____

   2 x 8 = _____

6. 11 x 11 = _____

   8 x 14 = _____

7. 13 x 13 = _____

   10 x 16 = _____

8. 15 x 15 = _____

   12 x 18 = _____

9. 18 x 18 = _324_

   15 x 21 = _____

10. 19 x 19 = _361_

    16 x 22 = _____

Now complete problems 11–20 using the pattern.
What is your discovery? Write it at the end.

11. 10 x 10 = _____

7 x 13 = _____

12. 12 x 12 = _144_

9 x 15 = _____

13. 20 x 20 = _400_

17 x 23 = _____

14. 30 x 30 = _900_

27 x 33 = _____

15. 40 x 40 = _1,600_

37 x 43 = _____

16. 50 x 50 = _2,500_

47 x 53 = _____

17. 60 x 60 = _3,600_

57 x 63 = _____

18. 70 x 70 = _____

67 x 73 = _____

19. 80 x 80 = _____

77 x 83 = _____

20. 90 x 90 = _____

87 x 93 = _____

Your Discovery: _____

_____

_____

Squares and Patterns IV

Complete problems 1–10 and look for a pattern.

Example:

1. 9 x 9   = ___81___

    5 x 13   = ___65___

2. 12 x 12 = ___144___

    8 x 16   = _____

3. 13 x 13 = ___169___

    9 x 17   = _____

4. 20 x 20 = ___400___

    16 x 24 = _____

5. 30 x 30 = ___900___

    26 x 34 = _____

6. 15 x 15 = ___225___

    11 x 19 = _____

7. 17 x 17 = ___289___

    13 x 21 = _____

8. 19 x 19 = ___361___

    15 x 23 = _____

9. 40 x 40 = ___1,600___

    36 x 44 = _____

10. 50 x 50 = ___2,500___

    46 x 54 = _____

Now complete problems 11–20 using the pattern.
What is your discovery? Write it at the end.

11. 60 x 60 = *3,600*

56 x 64 = _____

12. 70 x 70 = *4,900*

66 x 74 = _____

13. 80 x 80 = *6,400*

76 x 84 = _____

14. 90 x 90 = *8,100*

86 x 94 = _____

15. 100 x 100 = *10,000*

96 x 104 = _____

16. 200 x 200 = *40,000*

196 x 204 = _____

17. 10 x 10 = *100*

6 x 14 = _____

18. 25 x 25 = *625*

21 x 29 = _____

19. 16 x 16 = *256*

12 x 20 = _____

20. 18 x 18 = *324*

14 x 22 = _____

Your Discovery: _____

_____

_____

Squares and Patterns Mixed

Complete the following using discoveries you've made
in previous exercises. You should be able to do these
mentally.

	Problem:	**Answer:**	**Rule:**
1.	11 x 11 = 121		
	9 x 13 =	_117_	_____
2.	15 x 15 = 225		
	14 x 16 =	___	_____
3.	17 x 17 = 289		
	16 x 18 =	___	_____
4.	25 x 25 = 625		
	22 x 28 =	___	_____
5.	30 x 30 = 900		
	28 x 32 =	___	_____
6.	40 x 40 = 1,600		
	37 x 43 =	___	_____
7.	25 x 25 = 625		
	23 x 27 =	___	_____
8.	50 x 50 = 2,500		
	47 x 53 =	___	_____
9.	12 x 12 = 144		
	9 x 15 =	___	_____
10.	16 x 16 = 256		
	14 x 18 =	___	_____

Extra for experts:

11. $20 \times 20 = 400$

    $18 \times 22 =$ _____ _____

12. $30 \times 30 = 900$

    $26 \times 34 =$ _____ _____

13. $16 \times 16 = 256$

    $12 \times 20 =$ _____ _____

14. $15 \times 15 = 225$

    $11 \times 19 =$ _____ _____

15. $20 \times 20 = 400$

    $16 \times 24 =$ _____ _____

16. $30 \times 30 = 900$

    $27 \times 33 =$ _____ _____

17. $50 \times 50 = 2,500$

    $46 \times 54 =$ _____ _____

18. $60 \times 60 = 3,600$

    $56 \times 64 =$ _____ _____

19. $30 \times 30 = 900$

    $26 \times 34 =$ _____ _____

20. $100 \times 100 = 10,000$

    $98 \times 102 =$ _____ _____

Power of Ten I

Complete problems 1–10 and look for a pattern.

1. $10 \times 10 =$ _____

2. $100 \times 10 =$ _____

3. $100 \times 100 =$ _____

4. $100 \times 1{,}000 =$ _____

5. $1{,}000 \times 1{,}000 =$ _____

6. $10^1 \times 10^1 =$ _____

7. $10^1 \times 10^2 =$ _____

8. $10^2 \times 10^2 =$ _____

9. $10^2 \times 10^3 =$ _____

10. $10^3 \times 10^3 =$ _____

11. $10^3 \times 10^4 =$ _____

12. $10^4 \times 10^4 =$ _____

What pattern did you find? _____

Solve the following problems mentally.

13. $10 \times 20 =$ _____

14. $100 \times 20 =$ _____

15. $1{,}000 \times 20 =$ _____

16. $10 \times 30 =$ _____

17. $100 \times 30 =$ _____

18. $1{,}000 \times 30 =$ _____

19. $10 \times 200 =$ _____

20. $100 \times 200 =$ _____

21. $1000 \times 200 =$ _____

22. $10 \times 300 =$ _____

Power of Ten II

Complete the following as shown in the example.

Example:

1. $10 \times 20 = 10 \times 2 \,(10) = 10^1 \times 10^1 \times 2 = \underline{\quad 200 \quad}$

2. $100 \times 20 = 10 \times 10 \times 10 \times 2 = 10^3 \times 2 = \underline{\qquad\qquad}$

3. $1{,}000 \times 20 = 10 \times 10 \times 10 \times 10 \times 2 = 10^4 \times 2 = \underline{\qquad\qquad}$

4. $10 \times 30 = 10 \times 10 \times 3 = 10^2 \times 3 = \underline{\qquad\qquad}$

5. $100 \times 30 = 10 \times 10 \times 10 \times 3 = 10^3 \times 3 = \underline{\qquad\qquad}$

6. $1{,}000 \times 30 = 10 \times 10 \times 10 \times 10 \times 3 = 10^4 \times 3 = \underline{\qquad\qquad}$

7. $10 \times 200 = 10 \times 10 \times 10 \times 2 = 10^3 \times 2 = \underline{\qquad\qquad}$

8. $100 \times 200 = 10 \times 10 \times 10 \times 10 \times 2 = 10^4 \times 2 = \underline{\qquad\qquad}$

9. $1{,}000 \times 200 = 10 \times 10 \times 10 \times 10 \times 10 \times 2 = 10^5 \times 2 = \underline{\qquad\qquad}$

10. $100 \times 300 = 10 \times 10 \times 10 \times 10 \times 3 = 10^4 \times 3 = \underline{\qquad\qquad}$

11. $1{,}000 \times 300 = 10 \times 10 \times 10 \times 10 \times 10 \times 3 = 10^5 \times 3 = \underline{\qquad\qquad}$

12. $1{,}000 \times 400 = 10 \times 10 \times 10 \times 10 \times 10 \times 4 = 10^5 \times 4 = \underline{\qquad\qquad}$

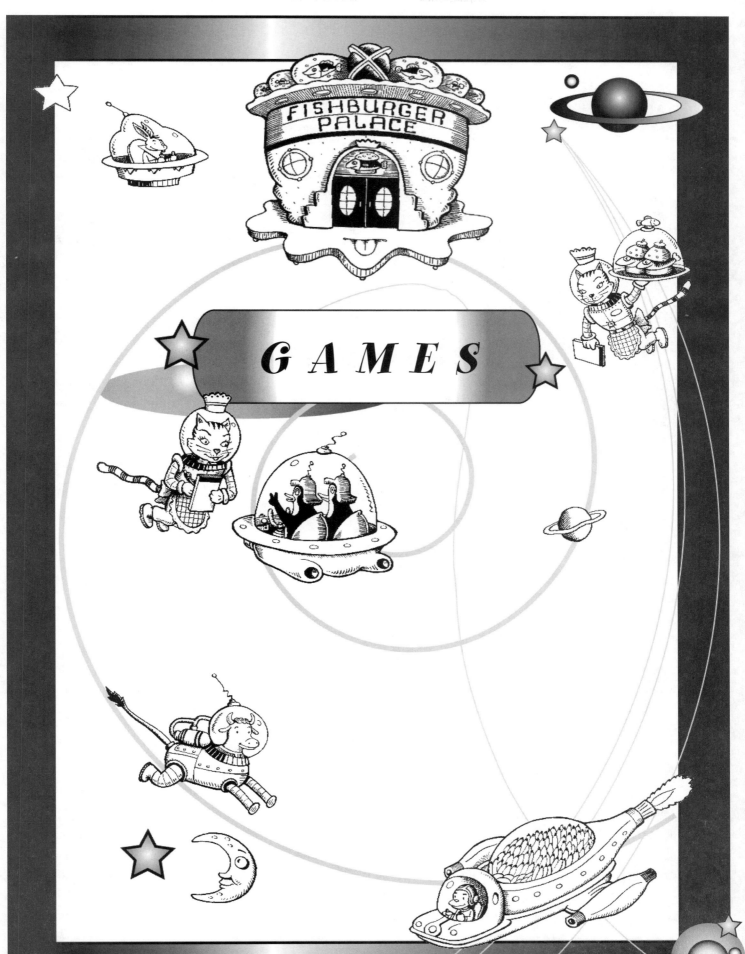

FISHBURGER PALACE

GAMES

Perfect Diamond

How to Play the Game

One or more players can play Perfect Diamond. Players use a die or a spinner with numbers 0–9. Spin to get the number and then write the number in one of the circles in the diamond.

Object of the Game

Players roll the die as many times as necessary to get three numbers in a straight line that equal 10 or 20. The player who completes the most lines wins.

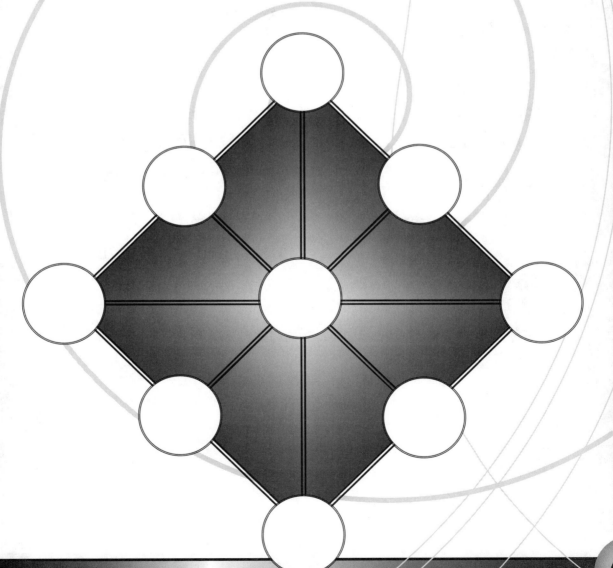

Greater Than or Less Than

How to Play the Game

Two players can play Greater Than or Less Than. Players use a die or a spinner with numbers 0–9. Player A rolls a die four times, and secretly records a number that can be obtained from the four numbers. Player B does the same thing. Neither player can see what the other rolls. Player A calls whether his or her number is larger or smaller than Player B's number.

Object of the Game

If Player A guesses correctly, he or she scores one point. If not, Player B scores one point. In the next set, players reverse the order.

Example Game:

1. **Player A** rolls 3, 1, 2, and 5.

2. **Player B** rolls 6, 9, 4, and 2.

3. **Player A** records 1, 2, 3, and 5.

4. **Player B** records 2, 4, 6, and 9.

5. **Player A** calls that his or her number is less.

6. Players show their numbers.

7. **Player A** scores 1 point.

8. Game continues for 10 rounds. Players add up their points and the player with the highest score wins.

Sums of 10

How to Play the Game

Two or more players can play Sums of 10. Players use a die with numbers 0–6. Players roll the die ten times and record the numbers on a sheet of paper. Players then choose all the sets of numbers within the set that equal 10 when added together. Each number can be used only once for each time it was rolled. Players score one point for each set of tens. Each player does this for a total of four rounds to complete a game. The player with the greatest score wins.

Object of the Game

The object of the game is to practice finding sums that yield compatible numbers, and, of course, to win!

Alternate Game 1

Players use a spinner with numbers 0–9 and try to make sums of 15.

Alternate Game 2

Players use a spinner with numbers 0–9 and try to make sums of 25.

Example Game:

1. **Player A** rolls 1, 3, 4, and 5, 4, 4, and 2.
   Sums: 6+4, 5+5, 4+4+2
   3 points

2. **Player B** rolls 3, 3, 4, 6, 6, 2, 2, 1, 4, and 6.
   Sums: 3+3+4, 6+4, 6+2+2
   3 points

3. **Player C** rolls 2, 2, 1, 3, 4, 2, 6, 6, 5, and 5.
   Sums: 2+2+1+3+2, 6+4, 5+5
   3 points

4. **Player D** rolls 4, 4, 6, 2, 1, 3, 4, 5, 5, and 6.
   Sums: 6+4, 5+5, 2+1+3+4
   3 points

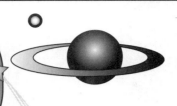

Target 1,000

How to Play the Game

Two or more players can play Target 1000.

Players use two dice (example colors: white and green) with numbers 0–9, or two spinners with numbers 0–9. If spinners are used, the first spinner is assigned a place value of ones and the second is assigned a place value of tens. If dice are used, the white die represents ones and the green represents tens.

Object of the Game

The object of the game is to get as close to 1,000 as possible without going over. Players roll or spin 15 times, each time rounding the number to the closest 50. For example, 24 or less will round to 0. The player who gets a total closest to 1,000 wins.

Alternate Game 1

Use three dice or three spinners with place values of 1, 10, and 100. Round to the nearest 100. Players spin or roll the dice 12 times or fewer. The object of the game is to reach a total of 10,000 without going over.

Alternate Game 2

In this variation, the players use two dice or spinners and multiply the first number shown by the second number shown. Round the product to the nearest ten. Try to reach the number 500 in 12 turns or fewer.

Target 1,000

Turn	Number	Rounded Number	Cumulative Total
1.			
2.			
3.			
4.			
5.			
6.			
7.			
8.			
9.			
10.			
11.			
12.			
13.			
14.			
15.			

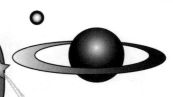

Guess My Number

• GAME 1 •

How to Play Game 1:

Two or more players can play Guess My Number. The following is an example of a game with two players.

On a piece of paper, **Player A** writes any whole number between 0 and 1000. **Player B** tries to guess the number in fewer than ten questions. The questions can only be answered by "yes" or "no."

Example Game 1:

Player A writes "465" secretly on a piece of paper.

1. **Player B** asks, "Is the number less than 500?"
   Player A answers, "Yes."

2. **Player B**: "Is the number less than 250?"
   Player A: "No."

3. **Player B**: "Is the number greater than 400?"
   Player A: "Yes."

4. **Player B**: "Is the number greater than 450?"
   Player A: "Yes."

5. **Player B**:"Is the number greater than 475?"
   Player A: "No."

6. **Player B**:"Is the number divisible by 5?"
   Player A: "Yes."

7. **Player B**: "Does the number end in a zero?"
   Player A: "No."

8. **Player B**: "Is the number 455?"
   Player A: "No."

9. **Player B**: "Is the number 465?"
   Player A: "Yes."

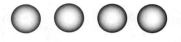

Guess My Number

• GAME 2 •

How to Play Game 2:

Two or more players can play Guess My Number. The following is an example of a game with two players.

On a piece of paper, **Player A** writes any fraction between 0 and

1. **Player B** tries to guess the number in fewer than ten questions. The questions can only be answered by "yes" or "no."

Example Game 2:

Player A writes "$\frac{2}{3}$" secretly on a piece of paper.

1. **Player B** asks, "Is the number less than $\frac{1}{2}$?"
   Player A answers, "No."

2. **Player B**: "Is the number a unit fraction?"
   Player A: "No."

3. **Player B**: "Is the number a terminating decimal?"
   Player A: "No."

4. **Player B**: "Is the denominator less than 5?"
   Player A: "Yes."

5. **Player B**: "Is the numerator 3?"
   Player A: "Yes."

6. **Player B**: "Is the number $\frac{2}{3}$?"
   Player A: "Yes."

Number Golf

How to Play the Game

Two or more players can play Number Golf. Players use a die or a spinner with numbers 0–9. Players roll the die four times and get four numbers that can be placed in the fraction format, "a/b + c/d." (The number 0 cannot be used for the denominator.)

Object of the Game

The par, or target score for each hole, matches the number of the hole. For example, at the first hole, players try to make a total as close to 1 as possible. At the second hole, players try to make a total as close to 2 as possible and, at the third hole, players try to make a total as close to 3 as possible.

The player with a total closest to par for a given hole scores a point. Since it is sometimes difficult to see which fraction sum is closest to the par, a calculator can be used. For example: First hole, Par 1.

Example Game:

Player A rolls 3, 2, 1, 6.

Player B rolls 9, 2, 2, 5.

Player A arranges his or her four numbers to form two fractions and then adds them together. Then he or she makes the fraction into a decimal by dividing the numerator by the denominator.

$$\frac{2}{3} + \frac{1}{6} = \frac{5}{6} = .833$$

Player B arranges his or her four numbers to form two fractions and then adds them together. Then he or she makes the fraction into a decimal by dividing the numerator by the denominator. Players are allowed to use a calculator.

$$\frac{2}{9} + \frac{2}{5} = \frac{10}{45} + \frac{18}{45} = \frac{28}{45} = .622$$

Since .833 is closer to 1 than .622, **Player A** scores 1 point.

The game continues for a total of nine holes. Players add up their points. The player with the highest score wins.

Fraction Golf

How to Play the Game

Two, three, or four players can play Fraction Golf.

Players use a die or a spinner with numbers 0–9. Players roll the die four times and get four numbers that can be used in the fraction format "a/b (+ or −) c/d." (The number 0 cannot be used for the denominator.)

Players then add or subtract these fractions together. If the total contains a fraction, the number must be rounded to the nearest whole number. Calculators may be used if needed.

For an added challenge, each hole can be timed at a maximum of two minutes.

Hole 1 Par 4
Hole 2 Par 4
Hole 3 Par 3
Hole 4 Par 5
Hole 5 Par 4
Hole 6 Par 4
Hole 7 Par 3
Hole 8 Par 5
Hole 9 Par 4

Object of the Game

The object of the game is to get as close to par as possible for each hole. If the player makes par, he or she receives one point for that hole. After the last hole the points are added up for each player and the player with the highest score wins. (In traditional golf, the player with the lowest score wins.) In the following examples, there are two players.

Example Game 1, Hole 1:

Player A rolls 2, 6, 1, 3 and writes:
$$\frac{6}{3} + \frac{2}{1} = 2 + 2 = 4$$

Player B rolls 3, 9, 2, 5 and writes:
$$\frac{9}{2} - \frac{3}{5} = \frac{45}{10} - \frac{6}{10} = 3.9 = 4$$

Since they both reach par, **Player A** and **Player B** each score one point for hole 1. Players proceed to play the remaining 8 holes and total their points for each hole.

Alternate Game 1, Hole 1:

There are no fractions in this version.

Player A rolls 6, 3, 4, 2 and writes:

$$(6 - 4) + (3 - 2) = 3$$

Player B rolls 9, 4, 2, 1 and writes:

$$(9 - 4) - (2 - 1) = 5 - 1 = 4$$

Since the total for **Player B** was 4 for this hole, he or she scores 1 point. The total for **Player A** was 3, so he or she scores 0. Players proceed to play the remaining 8 holes and total their points for each hole.

Alternate Game 2, Hole 1:

In this game, players may add, subtract, multiply, or divide with whole numbers or fractions.

Player A rolls 9, 9, 3, 6 and writes:

$$(9 + 9) \div (6 + 3) = 6$$

Player B rolls 3, 5, 8, 7 and writes:

$$(8 \div 3) + (7 \div 5) = 4$$

If **Player B** is using fractions, this could also be written as:

$$\frac{40}{15} + \frac{21}{15} = \frac{61}{15} = 4\frac{1}{15}$$
$$4\frac{1}{15} = 4 \text{ (rounded off)}$$

Player A and **Player B** each score 1 point because they both reached a par of 4.

How to Play the Game

Two or more players can play Millennium Madness. Players use a die or a spinner with numbers 0–6. Players roll the die five times and multiply the numbers by either 1, 10, or 1,000.

Object of the Game

Players try to reach a total of 1,000 without going over. The player with a total closest to 1,000 scores one point. The game repeats until one player reaches a score of 5 and wins the game.

In the example below, Player C wins one point because his total is closest to 1,000.

Player A	Player B	Player C
Rolls 6 Chooses 6 x 100 = 600 **600**	Rolls 5 Chooses 5 x 100 = 500 **500**	Rolls 5 Chooses 5 x 100 = 500 **500**
Rolls 3 Chooses 3 x 100 = 300 **900**	Rolls 6 Chooses 6 x 10 = 60 **560**	Rolls 2 Chooses 2 x 100 = 200 **700**
Rolls 4 Chooses 4 x 10 = 40 **940**	Rolls 4 Chooses 4 x 100 = 400 **960**	Rolls 3 Chooses 3 x 10 = 30 **730**
Rolls 6 Chooses 6 x 1 = 6 **946**	Rolls 5 Chooses 5 x 1 = 5 **965**	Rolls 2 Chooses 2 x 100 = 200 **930**
Rolls 3 Chooses 3 x 10 = 30 **976**	Rolls 5 Chooses 2 x 10 = 20 **985**	Rolls 6 Chooses 6 x 10 = 60 **990**

Spin to Win

How to Play the Game

Two or more players can play Spin to Win. Players use a die or a spinner with numbers 0–9.

The game consist of four rounds, each with a different objective. In Round 1, a player will try and make the largest number. Each player spins, in turn, and must write a number in one of the four boxes. Once a number is placed it cannot be moved. One number can be rejected in each round. The player who makes the largest four-digit number wins the round and gets the indicated number of points. A player may declare the round over if he or she can correctly explain why further turns would not affect the outcome.

Object of the Game

Spin to Win provides practice in place value, probability, and higher level thinking skills. Players will soon learn that they can gain an advantage by being aware of the chances of numbers occurring or repeating.

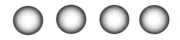

Player A	9	6	7		Reject	◇
Player B	9	8	4		Reject	◇

Example Game , Round 1:

Turn 1:

Player A spins a 6 and places it in the hundreds place.

Player B spins a 4 and places it in the tens place.

Turn 2:

Player A spins a 9 and places it in the thousands place.

Player B spins an 8 and places it in the hundreds place.

Turn 3:

Player A spins a 7 and places it in the tens place.

Player B spins a 9 and places it in the thousands place.

Game over.

When a player declares the game over, he or she must explain why. In this game, the greatest four-digit number wins. Therefore, the game was over when **Player B** put the 9 in the thousands place because no number in the ones place could change the game.

Spin to Win, Round 1:

Make the largest four-digit number.

Player A  Reject ◇

Player B  Reject ◇

Player A points: ____

Player B points: ____

Spin to Win, Round 2:

Make the smallest four-digit number.

Player A   Reject ◇

Player B   Reject ◇

Player A points: ____

Player B points: ____

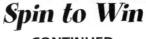

Spin to Win
• CONTINUED •

Spin to Win, Round 3:

Make a four-digit number as close to 1,000 as possible (6 points).

Player A ☐ ☐ ☐ ☐   Reject ◇

Player B ☐ ☐ ☐ ☐   Reject ◇

Player A points: _____

Player B points: _____

Spin to Win, Round 4:

Make a four-digit number as close to 5,000 as possible (6 points).

Player A ☐ ☐ ☐ ☐   Reject ◇

Player B ☐ ☐ ☐ ☐   Reject ◇

Player A points: _____

Player B points: _____

Game Total:

Player A points: _____

Player B points: _____

Answers

Pages 2 and 3

2. 21, 24, 27, 30, 33. Add 3.

3. 38, 43, 48. Add 5.

4. 40, 47, 54, 61. Add 7.

5. 49, 64, 81. They are the squares of consecutive numbers.

6. 32, 29, 26, 23. Subtract 3.

7. 37, 34, 36, 33. Add 2, subtract 3.

8. 256, 1,024, 4,096. Multiply by 4.

10. 4, $4\frac{1}{2}$, 5, $5\frac{1}{2}$. Add $\frac{1}{2}$.

11. $\frac{1}{32}$, $\frac{1}{64}$, $\frac{1}{128}$. Multiply by $\frac{1}{2}$.

12. 128, 256, 512, 1,024. Double the number.

13. 2, 3, 13, 17, 23. Prime numbers in sequence.

14. 3, 9, 27. Multiply by 3.

15. 18, 23, 28, 33. Add 5.

16. 6, $\frac{1}{6}$, 7, $\frac{1}{7}$. The last whole number in the sequence becomes the denominator in the fraction that follows it. The numerator is always 1.

17. 1, 2, 1. Keep adding another 1 to the sequence.

18. 13, 21. Each new number is the sum of the previous two numbers.

19. 13, 15. Add 2 to the last number in the sequence.

20. 1,936, 3,872. Double the number.

Pages 4 and 5

2. 6, 8. (Repeats 2, 4, 6, 8.)

3. 11, 13, 15. (Add 2.)

4. 26, 32, 38. (Add 6.)

5. 15, 18, 21, 24. (Add 3.)

6. 25, 30, 35. (Add 5.)

7. 30, 36, 42. (Add 6.)

8. 21, 24, 27. (Add 3.)

9. 2, 4, 6. (Repeats 2, 4, 6.)

10. 3, 9, 9. (Ascending: 1, 3, 9, Descending: 9, 3, 1, etc.)

11. 18, 23, 28, 33 (Add 5.)

12. 4, $4\frac{1}{2}$, 5, $5\frac{1}{2}$. (Add $\frac{1}{2}$.)

13. 80, 75, 70. (Subtract 5.)

14. 1, 2, 3, 4, 5, 6, 7, 8. (Consecutive numbers.)

15. 1, 3, 7, 13, 21, 31. (Add 2, 4, 6, 8, 10.)

16. $\frac{5}{3}$, $\frac{6}{3}$, $\frac{7}{3}$.(Numerators are consecutive numbers.)

17. $\frac{12}{15}$, $\frac{15}{18}$, $\frac{18}{21}$. (Numerators and denominators increase by 3 each time.)

18. 16, 32, 64. (Double each number.)

19. 0, 15. (Add 15 starting with 0.)

20. 243, 729, 2,187. (3^1, 3^2, 3^3, 3^4, 3^5.)

Page 6

Answers will vary but here are some possibilities:

1. 9 is the only one-digit number.

2. 35 is out of sequence. Others differ by 2.

3. 47. Others divide by 2.

4. 9 is out of sequence.

5. $\frac{2}{5}$ is the only number that is not a unit fraction.

6. 25 is the only odd number.

7. 954 is not in the pattern.

8. 235 is not a perfect square.

9. 10 is not an odd number.

10. 37. Others divide by 3.

Page 7

Answers will vary but here are some possibilities.

1. 5 is a prime number.

2. 16 is out of sequence.

3. 60 is out of sequence.

4. 4 is duplicated.

5. $\frac{2}{3}$ is not a unit fraction.

6. 3.0 is the only number >1.

7. 357 is the only odd number.

8. 47 is not a square number.

9. 1 is the only one-digit number.

10. 171 is the only three-digit number.

Page 8

Answers will vary but here are some possibilities.

2. $\frac{4}{6}$ has not been reduced—the others have.

3. 6% is the only single digit number.

4. $\frac{2}{3}$ is greater than $\frac{1}{2}$.

5. $\frac{8}{7}$ is greater than 1.

6. $\frac{1}{3}$ is the only number that is not equal to $\frac{1}{2}$.

7. 2.0 is the only number that is not equal to the others.

8. $\frac{12}{15}$ is the only number that is not equal to the others.

9. 123% is the only number that is greater than 100%.

10. $\frac{2}{5}$ is the only number that is less than $\frac{1}{2}$.

Page 9

Answers may vary but here are some possibilities.

2. 6.8 is the only number that is greater than 1.

3. 27% is not a multiple of 5.

4. $\frac{2}{5}$ is the only number that does not have 3 in the denominator.

5. $\frac{9}{10}$ is the only number that has a two-digit denominator.

6. $\frac{6}{12}$ is the only number that does not equal $\frac{1}{3}$.

7. .30 is the only number that is not expressed as a fraction.

8. $\frac{16}{25}$ is the only number that is not equal to $\frac{3}{5}$.

9. 153% is the only number that reverses the order of 1, 3, 5.

10. $\frac{8}{7}$ is the only number that is greater than 1.

Page 10

Answers will vary but here are some possibilities.

1. They are all divisible by 2.
2. They are all divisible by 3.
3. They are all odd numbers.
4. They are all even numbers.
5. They are all divisible by 5.
6. They are all prime numbers.
7. They are all all palindromes.
8. They are all divisible by 4.
9. The sum of the digits equals 10.
10. The sum of the digits equals 9.

Page 11

Answers will vary but here are some possibilities.

1. They are all composite numbers.
2. They are all divisible by 3.
3. They are all divisible by 4.
4. They are all divisible by 7.
5. They are all even numbers.
6. They are all three-digit numbers.
7. They are all divisible by 4.
8. They are all between 50 and 59.
9. They are divisible by 3.
10. They are all prime numbers.

Page 12

Answers will vary but here are some possibilities:

2. They are all perfect cubes.
3. They are all less than or equal to $\frac{1}{2}$.
4. They are all odd numbers.
5. They are all even numbers.
6. They are all even numbers.
7. They are all prime numbers.
8. They are all less than 20.
9. They are less than 15.
10. They are all less than 22.

Page 13

Answers will vary but here are some possibilities:

2. They are all divisible by 3.
3. They are all two-digit numbers.
4. The sum of their digits is less than 10.
5. The sum of their digits is greater than 10.
6. They are all composite numbers.
7. They are all less than 70.
8. They are all less than 40.
9. They are all two-digit numbers.
10. None are divisible by 6.

Answers

Page 14

	Thous.	Hun.	Tens	Units
2.	3	1	0	2
3.	0	0	5	7
4.	0	6	0	0
5.	5	0	1	0
6.	1	6	8	4
7.	0	5	2	3
8.	1	0	1	0
9.	3	3	0	3
10.	0	7	0	7

Page 15

	Thous.	Hun.	Tens	Units
2.	5	4	0	7
3.	1	0	0	5
4.	3	6	1	0
5.	2	9	3	8
6.	3	0	0	7
7.	4	5	0	6
8.	3	0	7	6
9.	4	9	0	3
10.	6	0	1	0

Page 16

2. 3,056

3. 3,406

4. 3,450

5. 456

6. 1,006

7. 1,450

8. 3,407

9. 1,060

10. 101

Page 17

2. 707

3. 955

4. 1,201

5. 11,026

6. 733

7. 3,936

8. 291

9. 7,625

10. 2,089

Page 18

1. 10

2. 10

3. 0

4. 40

5. 100

6. 200

7. 300

8. 1

9. 1

10. 1

11. 0

12. 0

13. 0

14. 1

15. 7

16. 5

17. 4,000

18. 4,000

19. 9,000

20. 4,000

21. 3,000

22. 6,000

Page 19

2. 300

3. 200

4. 700

5. 800

6. 4,000

7. 4,000

8. 9,000

9. 4,000

10. 3,000

11. 0

12. 1

13. 7

14. 5

15. 8

Page 20

2.	50	9.	800
3.	130	10.	1,000
4.	260	11.	1
5.	670	12.	1
6.	200	13.	1
7.	400	14.	0
8.	800	15.	8

Page 21

2.	200	9.	3,000
3.	300	10.	1,000
4.	500	11.	0
5.	700	12.	1
6.	4,000	13.	7
7.	4,000	14.	5
8.	9,000	15.	7

Page 22

Answers may vary.

2.	40 + 120	160
3.	60 + 70	130
4.	70 + 100	170
5.	500 – 50	450
6.	700 – 400	300
7.	400 – 200	200
8.	500 – 300	200
9.	60 x 20	1,200
10.	40 x 40	1,600
11.	100 x 40	4,000
12.	100 x 200	20,000

13.	300 ÷ 20	15
14.	500 ÷ 30 (approx.)	16
15.	900 ÷ 50 (approx.)	18
16.	1,400 ÷ 60 (approx.)	23

Page 23

Answers may vary.

2.	80 + 40	120
3.	40 + 40	80
4.	100 + 70	170
5.	560 – 40	520
6.	750 – 320	430
7.	120 – 50	70
8.	360 – 40	320
9.	50 x 40	2,000
10.	40 x 40	1,600
11.	30 x 30	900
12.	100 x 150	15,000
13.	36 ÷ 6	6
14.	60 ÷ 4	15
15.	120 ÷ 20	60
16.	1,400 ÷ 30 (approx.)	40

Page 24

Answers may vary.

2.	40 + 70	110
3.	120 + 230	350
4.	250 + 350	600
5.	100 – 40	60
6.	70 – 40	300
7.	140 – 40	100

8. 250 − 150	100	
9. 20 x 20	400	
10. 30 x 60	1,800	
11. 80 x 60	4,800	
12. 100 x 40	4,000	
13. 49 ÷ 7	7	
14. 100 ÷ 10	10	
15. 130 ÷ 13	10	
16. 2,250 ÷ 150	15	

Page 25

Answers may vary.

2. 20 ÷ 4	5
3. 30 ÷ 6	5
4. 130 ÷ 30 (approx.)	4
5. 100 + 40	140
6. 160 + 40	200
7. 60 + 40	100
8. 400 + 130	530
9. 10 x 13	130
10. 10 x 21	210
11. 15 x 15	225
12. 100 x 40	4,000
13. 120 ÷ 60	2
14. 120 ÷ 4	30
15. 1,200 ÷ 6	200
16. 1,400 ÷ 70	20

Page 26

1. 6	**14.** 8
2. 5	**15.** 6
3. 5	**16.** 2
4. 0	**17.** 6
5. 1	**18.** 7
6. 3	**19.** 5
7. 4	**20.** 3
8. 9	**21.** 0
9. 2	**22.** 4
10. 9	**23.** 6
11. 8	**24.** 2
12. 6	**25.** 4
13. 0	

Page 27

1. 5	**14.** 7
2. 9	**15.** 4
3. 3	**16.** 4
4. 2	**17.** 1
5. 3	**18.** 2
6. 4	**19.** 4
7. 9	**20.** 6
8. 8	**21.** 4
9. 8	**22.** 4
10. 9	**23.** 8
11. 2	**24.** 2
12. 8	**25.** 4
13. 6	

Pages 28 and 29

Answers may vary.

2. $70 + 130 = 200$ 2
3. $130 + 360 = 490$ 1
4. $950 + 350 = 1,300$ 5
5. $1,300 + 2,400 = 3,700$ 5
6. $4,000 - 2,400 = 1,600$ 3
7. $5,000 - 3,500 = 1,500$ 4
8. $9,000 - 4,000 = 5,000$ 9
9. $8,000 - 4,000 = 4,000$ 2
10. $10,000 - 9,000 = 1,000$ 4
11. $20 \times 15 = 300$ 2
12. $20 \times 20 = 400$ 9
13. $40 \times 60 = 2,400$ 2
14. $60 \times 90 = 5,400$ 2
15. $200 \times 200 = 4,000$ 4
16. $1,400 \times 200 = 28,000$ 6
17. $200 \times 40 = 8,000$ 2
18. $700 \times 20 = 14,000$ 4
19. $100 \times 32 = 3,200$ 4
20. $140 \times 60 = 8,400$ 4
21. $90 \times 90 = 8,100$ 4
22. $50 + 90 = 140$ 5
23. $80 \times 140 = 11,200$ 3
24. $40 \times 50 = 2,000$ 2
25. $80 \times 40 = 3,200$ 0
26. $20 + 20 = 40$ 6
27. $50 - 16 = 34$ 2
28. $25 \times 25 = 625$ 0
29. $40 \times 20 = 800$ 8

30. $15 \times 15 = 225$ 6
31. $40 - 20 = 20$ 0
32. $160 \div 2 = 80$ 9
33. $1,200 \div 3 = 400$ 1
34. $1,600 \div 4 = 400$ 4
35. $5,600 \div 8 = 700$ 1

Page 30

Answers may vary.

2. $20 + 40 = 60$ 0
3. $30 + 50 = 80$ 1
4. $140 + 60 = 200$ 2
5. $240 + 340 = 580$ 4
6. $5,000 - 3,000 = 2,000$ 6
7. $3,600 - 3,000 = 600$ 2
8. $7,000 - 5,000 = 2,000$ 2
9. $8,000 - 6,300 = 1,700$ 0
10. $10,000 - 5,000 = 5,000$ 2
11. $15 \times 15 = 225$ 0
12. $20 \times 50 = 1,000$ 1
13. $40 \times 30 = 1,200$ 8
14. $60 \times 60 = 3,600$ 8
15. $1,400 \times 60 = 84,000$ 4
16. $1,400 \times 50 = 7,000$ 8
17. $180 \times 40 = 7,200$ 8
18. $70 \times 50 = 3,500$ 4
19. $160 \times 20 = 3,200$ 3
20. $400 \times 15 = 6,000$ 2
21. $100 \times 60 = 6,000$ 8
22. $50 \times 60 = 3,000$ 0

23. $40 + 140 = 180$ 4

24. $60 + 40 = 100$ 3

25. $140 + 40 = 180$ 4

26. $160 \div 8 = 20$ 1

27. $210 \div 7 = 30$ between 4 and 5

28. $360 \div 4 = 90$ between 5 and 6

29. $400 \div 5 = 80$ between 7 and 8

30. $1,600 \div 10 = 160$ 3

31. $400 \div 5 = 80$ between 6 and 7

32. $630 \div 7 = 90$ 0

33. $120 \div 10 = 12$ 3

34. $300 \div 6 = 50$ between 5 and 6

35. $4,000 \div 8 = 500$ 4

Pages 32 and 33

Answers may vary.

2. Estimate:
$140 + 260 = 400$

First solution:
$130 + 250 + 7 + 9 =$
$380 + 16 = 396$

Second solution:
$140 + 260 - 3 - 1 =$
$400 - 4 = 396$

3. Estimate:
$70 + 80 = 150$

First solution:
$70 + 75 - 2 =$
$145 - 2 = 143$

Second solution:
$60 + 70 + 8 + 5 =$
$130 + 13 = 143$

4. Estimate:
$40 + 40 = 80$

First solution:
$30 + 40 + 8 + 2 =$
$70 + 10 = 80$

Second solution:
$40 - 2 + 40 + 2 = 80$

Pages 34 and 35

Answers may vary.

2. Estimate:
$120 \div 6 = 20$

First solution:
$\frac{120}{6} + \frac{18}{6} = 20 + 3 = 23$

Second solution:
$6 \times 20 = 120$
$138 - 120 = 18$
$3 \times 6 = 18$
$20 = 3 = 23$

3. Estimate:
$250 \div 10 = 25$

First solution:
$\frac{210}{7} + \frac{39}{7} =$
$30 + 5\frac{4}{7} = 35\frac{4}{7}$

Second solution:
$\frac{210}{7} + \frac{35}{7} + \frac{4}{7} =$
$30 + 5 + \frac{4}{7} = 35\frac{4}{7}$

4. Estimate:
$130 \times 60 = 7,800$

First solution:
$130(60 + 4) = 7,800 + 520$
$= 8,320$

Second solution:
$(100 + 30)(60 + 4) =$
$6,000 + 1,800 + 400 + 120 =$
$7,800 + 520 = 8,320$

Pages 36

Solutions will vary. Answers are:

2. 527
3. 272
4. 169
5. 980
6. 2,496
7. 36
8. 55
9. 20
10. 26

Pages 37

Solutions will vary. Answers are:

2. 120
3. 506
4. 213
5. 1,050
6. 1,050
7. $8\frac{1}{2}$
8. $20\frac{2}{3}$
9. 46
10. 21

Pages 38 and 39

2. No
3. No
4. Yes
5. No
6. Yes
7. Yes
8. No
9. Yes (if one is negative)
10. No
11. Yes
12. Yes
13. Yes
14. Yes
15. Yes (if one is negative)
16. No
17. Yes
18. No
19. Yes
20. No
21. No
22. Yes
23. Yes
24. No
25. Yes

Pages 40 and 41

2. No
3. No
4. Yes
5. Yes
6. No
7. Yes
8. Yes
9. No
10. No
11. No
12. No
13. No
14. Yes
15. Yes
16. Yes
17. Yes
18. Yes
19. Yes
20. Yes
21. No
22. No
23. Yes
24. No
25. No
26. No
27. Yes
28. No
29. No

Page 42

2. Yes
3. Yes
4. Yes
5. Yes

Page 43

1. Yes
2. No
3. Yes
4. Yes
5. Yes

Page 44

1. No
2. Yes
3. Yes
4. Yes
5. Yes

Page 45

1. Yes
2. Yes
3. No
4. Yes
5. Yes

Page 46

1. U
2. R
3. U
4. R
5. R

Page 47

1. R
2. U
3. U
4. U
5. R

Page 48

1. Yes (but incorrect)
2. No
3. No

4. No
5. No

Page 49

1. Yes
2. Yes
3. No
4. Yes
5. No

Page 50

1. No
2. No
3. No
4. Yes
5. No

Page 51

1. No
2. No
3. No
4. Yes
5. No

Page 52

What kind of numbers were in problems 1–8? Even numbers

What kind of numbers were in problems 9–16? Odd numbers

What kind of numbers were in problems 17–24? Odd and even numbers

What always happens when you add two even numbers? You get an even number.

What always happens when you add two odd numbers? You get an even number.

What always happens when you add an even and an odd number? You get an odd number.

Page 53

What kind of numbers were in problems 1–8? Even numbers

What kind of numbers were in problems 9–16? Odd numbers

What kind of numbers were in problems 17–24? Odd and even numbers

What always happens when you multiply two even numbers? You get an even number.

What always happens when you multiply two odd numbers? You get an odd number.

What always happens when you multiply an even and an odd number? You get an even number.

Page 54

What kind of numbers were in problems 1–8? Even minus even numbers.

What kind of numbers were in problems 9–16? Odd minus even numbers.

What kind of numbers were in problems 17–24? Even numbers minus odd numbers.

What always happens when you subtract an even number from another even number? You get an even number.

What always happens when you subtract an odd number from another odd number? You get an even number.

What always happens when you subtract an even number from an odd number? You get an odd number.

Page 55

What did you notice? No patterns exist.

Pages 56 and 57

Answers will vary but some sample answers are:

1. $30 + 35 = 65$
2. $45 + 35 + 4 = 80 + 4 = 84$
3. $50 + 40 + 6 + 1 = 97$
4. $95 + 65 - 2 = 160 - 2 = 158$
5. $80 + 60 + 1 = 141$
6. $80 + 60 - 2 = 140 - 2 = 138$
7. $65 + 65 + 2 = 132$
8. $90 + 80 - 2 = 168$
9. $123 + 43 + 4 = 166 + 4 = 170$
10. $130 + 55 = 185$
11. $238 + 70 - 1 = 308 - 1 = 307$
12. $1,200 + 2,900 + 46 + 38 = 4,100 + 40 + 44 = 4,184$
13. $639 + 999 = 638 + 1,000 = 1,638$
14. $234 + 567 = 230 + 560 + 4 + 7 = 790 + 11 = 801$
15. $345 + 786 = 340 + 780 + 5 + 6 = 1,120 + 11 = 1,131$

Pages 58 and 59

Answers will vary.

1. $8 (25) = 8 (20 + 5) = 160 + 40 = 200$
   $25 (10 - 2) = 250 - 50 = 200$

2. $6 (48) = 6 (40 + 8) = 240 + 48 = 288$
   $6 (50 - 2) = 300 - 12 = 288$

3. $9 (11) = 9 (10 + 1) = 90 + 9 = 99$
   $11 (10 - 1) = 110 - 11 = 99$

4. $7 (16) = 7 (10 + 6) = 70 + 42 = 112$
   $7 (20 - 4) = 140 - 28 = 112$

5. $10 (14) = 10 (10 + 4) = 100 + 40 = 140$
   $14 (10 + 0) = 140$

6. $10 (65) = 10 (60 + 5) = 600 + 50 = 650$
   $10 (65) = 650$

7. $12 (15) = 12 (10 + 5) = 120 + 60 = 180$
   $15 (10 + 2) = 150 + 30 = 80$

8. $15 (25) = 15 (20 + 5) = 300 + 75 = 375$
   $25 (10 + 5) = 250 + 125 = 375$

9. $3 (64) = 3 (60 + 4) = 180 + 12 = 192$
   $3 (70 - 6) = 210 - 18 = 192$

10. $5 (75) = 5 (70 + 5) = 350 + 25 = 375$
    $5 (80 - 5) = 400 - 25 = 375$

11. $5 (92) = 5 (100 - 8) = 500 - 40 = 460$
    $5 (90 + 2) = 450 + 10 = 460$

12. $9 (91) = 9 (100 - 9) = 900 - 81 = 819$
    $9 (90 + 1) = 810 + 9 = 819$

13. $8 (64) = 8 (60 + 4) = 450 + 32 = 512$
    $8 (70 - 6) = 560 - 48 = 512$

14. $9 (72) = 9 (70 + 2) = 630 + 18 = 648$
    $72 (10 - 1) = 720 - 72 = 648$

15. $13 (13) = 13 (10 + 3) = 130 + 39 = 169$
    $13 (11 + 2) = 13 (11) + 13 (2) =$
    $143 + 26 = 169$

Pages 60 and 61

Answers will vary.

1. $4 (24) = 4 (20) + 4 (4) = 80 + 16 = 96$
   $4 (25 - 1) = 100 - 4 = 96$

2. $3 (27) = 3 (20 + 7) = 60 + 21 = 81$
   $3 (30 - 3) = 90 - 9 = 81$

3. $4 (28) = 4 (25 + 3) = 100 + 12 = 112$
   $4 (20 + 8) = 80 + 32 = 112$

4. $5 (26) = 5 (20 + 6) = 100 + 30 = 130$
   $5 (25 + 1) = 125 + 5 = 130$

5. $11 (14) = 11 (10 + 4) = 110 + 44 = 154$
   $14 (10 + 1) = 140 + 14 = 154$

6. $12 (13) = 12 (10 + 3) = 120 + 36 = 156$
   $13 (10 + 2) = 130 + 26 = 156$

7. $6 (19) = 6 (20 - 1) = 120 - 6 = 114$
   $6 (10 + 9) = 60 + 54 = 114$

8. $12 (25) = 12 (20 + 5) = 240 + 60 = 300$
   $25 (10 + 2) = 250 + 50 = 300$

9. $9 (16) = 9 (10 + 6) = 90 + 54 = 144$
   $16 (10 - 1) = 160 - 16 = 144$

10. $8 (17) = 8 (10 + 7) = 80 + 56 = 136$
    $8 (20 - 3) = 160 - 24 = 136$

11. $5 (25) = 5 (20 + 5) = 100 + 25 = 125$
    $5 (30 - 5) = 150 - 25 = 125$

12. $6 (38) = 6 (40 - 2) = 240 - 12 = 228$
    $60 (30 + 8) = 180 + 48 = 228$

13. $9 (103) = 9 (100 + 3) = 900 + 27 = 927$
    $103 (10 - 1) = 1,030 - 103 = 927$

14. $9 (63) = 9 (60 + 3) = 540 + 27 = 567$
    $63 (10 - 1) = 630 - 63 = 567$

15. $12 (47) = 12 (50 - 3) = 600 - 36 = 564$
    $12 (40 + 7) = 480 + 84 = 564$

Pages 62 and 63

Answers will vary.

1. $5(25) = 5(20+5) = 100 + 25 = 125$
   $5(30-5) = 150 - 25 = 125$

2. $6(14) = 6(10+4) = 60 + 24 = 84$
   $6(20-6) = 120 - 36 = 84$

3. $7(22) = 7(20+2) = 140 + 14 = 154$
   $7(25-3) = 175 - 21 = 154$

4. $8(31) = 8(30+1) = 240 + 8 = 248$
   $8(40-9) = 320 - 72 = 248$

5. $9(27) = 9(20+7) = 180 + 63 = 243$
   $9(30-3) = 270 - 27 = 243$

6. $4(96) = 4(90+6) = 360 + 24 = 384$
   $4(100-4) = 400 - 16 = 384$

7. $5(32) = 5(30+2) = 150 + 10 = 160$
   $5(25+7) = 125 + 35 = 160$

8. $7(28) = 7(30-2) = 210 - 14 = 196$
   $7(20+8) = 140 + 56 = 196$

9. $6(43) = 6(40+3) = 240 + 18 = 258$
   $6(50-7) = 300 - 42 = 258$

10. $7(39) = 7(30+9) = 210 + 63 = 273$
    $7(40-1) = 280 - 7 = 273$

11. $8(62) = 8(60+2) = 480 + 16 = 496$
    $8(70-8) = 560 - 64 = 496$

12. $7(49) = 7(50-1) = 350 - 7 = 343$
    $7(40+9) = 280 + 63 = 343$

13. $9(96) = 9(100-4) = 900 - 36 = 864$
    $9(90+6) = 810 + 54 = 864$

14. $12(13) = 12(10+3) = 120 + 36 = 156$
    $13(10+2) = 130 + 26 = 156$

15. $16(18) = 16(10+10-2) =$
    $160 + 160 - 32 = 320 - 32 = 288$
    $18(10+10-4) =$
    $180 + 180 - 72 = 360 - 72 = 288$

Pages 64 and 65

Answers will vary.

1. $7(124) = 7(100+20+4) =$
   $700 + 140 + 28 = 868$
   $7(120+4) = 840 + 28 = 868$

2. $12(25) = 12(20+5) = 240 + 60 = 300$
   $25(10+2) = 250 + 50 = 300$

3. $25(63) = 25(60+3) = 1{,}500 + 75 = 1{,}575$
   $63(20+5) = 1{,}260 + 315 = 1{,}575$

4. $37(41) = 37(40+1) = 1{,}480 + 37 = 1{,}517$
   $41(30+7) = 1{,}230 + 287 = 1{,}517$

5. $64(29) = 64(20+9) = 1{,}280 + 576 = 1{,}856$
   $64(30-1) = 1{,}920 - 64 = 1{,}856$

6. $25(63) = 25(60+3) = 1{,}500 + 75 = 1{,}575$
   $(20+5)(60+3) =$
   $1{,}200 + 300 + 60 + 15 = 1{,}575$

7. $64(29) = 64(30-1) = 1{,}920 - 64 = 1{,}856$
   $(60+4)(20+9) =$
   $1{,}200 + 80 + 540 + 36 = 1{,}856$

8. $39(37) = (40-1)(40-3) =$
   $1{,}600 - 40 - 120 + 3 = 1{,}443$
   $37(40-1) = 1{,}480 - 37 = 1{,}443$

9. $16(16) = 16(10+6) = 160 + 96 = 256$
   $16(20-4) = 320 - 64 = 256$

10. $19(21) = 19(20+1) = 380 + 19 = 399$
    $21(10+9) = 210 + 189 = 399$

11. $32(34) = 32(30+4) = 960 + 128 = 1{,}088$
    $34(30+2) = 1{,}020 + 68 = 1{,}088$

12. $126(51) = 51(100+20+6) =$
    $5{,}100 + 1{,}020 + 306 = 6{,}426$
    $126(50+1) = 6{,}300 + 126 = 6{,}426$

13. $250 \, (62) = 250 \, (20 + 20 + 20 + 2) =$
$5{,}000 + 5{,}000 + 5{,}000 + 500 =$
$15{,}500$
$250 \, (60 + 2) = 15{,}000 + 500 =$
$15{,}500$

14. $314 \, (48) = (50 - 2) \, (300 + 14) =$
$15{,}000 - 600 + 700 - 28 =$
$15{,}072$
$48 \, (300 + 14) =$
$14{,}400 + 672 = 15{,}072$

15. $16 \, (18) = 16 \, (20 - 2) = 320 - 32 = 288$
$16 \, (10 + 8) = 160 + 128 = 288$

Pages 66 and 67

1. 81, 80
2. 49, 48
3. 25, 24
4. 100, 99
5. 121, 120
6. 16, 15
7. 36, 35
8. 169, 168
9. 289 ,288
10. 400, 399
11. 900, 899
12. 1,600, 1,599
13. 2,500, 2,499
14. 3,600, 3,599
15. 4,900, 4,899
16. 324, 323
17. 529, 528
18. 1,296, 1,295
19. 10,404, 10,403
20. 9,801, 9,800

Your Discovery: $a \times a = a^2$
$(a - 1) \, (a+1) = a^2 - 1$

When you multiply two numbers where the first number is 1 less than the second number, the product equals the first number squared minus 1.

Pages 68 and 69

2. 25, 21
3. 36, 32
4. 49, 45
5. 64, 60
6. 81, 77
7. 100, 96
8. 121, 117
9. 144, 140
10. 225, 221
11. 400, 396
12. 900, 896
13. 1,600, 1,596
14. 2,500, 2,496
15. 3,600, 3,596
16. 4,900, 4,896
17. 6,400, 6,396
18. 8,100, 8,096
19. 10,000, 9,996
20. 11,449, 11,445

Your Discovery: $(a - 2) \, (a + 2) = a^2 - 4$
When you multiply two numbers where the first number is 2 less than the second number, the product equals the first number squared minus 4.

Pages 70 and 71

1. 81, 72
2. 64, 55
3. 49, 40
4. 36, 27
5. 25, 16
6. 121, 112
7. 169, 160
8. 225, 216
9. 324, 315
10. 361, 352
11. 100, 91
12. 144, 135
13. 400, 391
14. 900, 891
15. 1,600, 1,591

16. 2,500, 2,491
17. 3,600, 3,591
18. 4,900, 4,891
19. 6,400, 6,391
20. 8,100, 8,091

Your Discovery: $(a - 3)(a + 3) = a^2 - 9$
When you multiply two numbers where the first number is 3 less than the second number, the product equals the first number squared minus 9.

Pages 72 and 73

1. 65
2. 128
3. 153
4. 384
5. 884
6. 209
7. 273
8. 345
9. 1,584
10. 2,484
11. 3,584
12. 4,884
13. 6,384
14. 8,084
15. 9,984
16. 39,984
17. 84
18. 609
19. 337
20. 308

Your Discovery: $(a - 4)(a + 4) = a^2 - 16$
When you multiply two numbers where the first number is 4 less than the second number, the product equals the first number squared minus 16.

Pages 74 and 75

Answer	Rule
1. 117	4 less than square
2. 224	1 less than square
3. 288	1 less than square
4. 616	9 less than square
5. 896	4 less than square
6. 1,591	9 less than square
7. 621	4 less than square
8. 2,491	9 less than square
9. 135	9 less than square
10. 252	4 less than square

Extra for experts:

11. 384	16 less than square
12. 884	16 less than square
13. 240	16 less than square
14. 209	16 less than square
15. 384	16 less than square
16. 891	9 less than square
17. 2,484	16 less than square
18. 3,584	16 less than square
19. 884	16 less than square
20. 9,996	4 less than square

Page 76

1. 100
2. 1,000
3. 10,000
4. 100,000
5. 1,000,000
6. $10^2 = 100$
7. $10^3 = 1,000$
8. $10^4 = 10,000$
9. $10^5 = 100,000$
10. $10^6 = 1,000,000$
11. $10^7 = 10,000,000$
12. $10^8 = 100,000,000$

What pattern did you find? By multiplying the first number and then adding the number of zeros, you can find the product.

13. 200
14. 2,000
15. 20,000
16. 300
17. 3,000
18. 30,000
19. 2,000
20. 20,000
21. 200,000
22. 3,000
23. 30,000
24. 300,000
25. 400,000

Page 77

2. 2,000
3. 20,000
4. 300
5. 3,000
6. 30,000
7. 2,000
8. 20,000
9. 200,000
10. 3,000
11. 300,000
12. 400,000